Contribution of X-ray Fluorescence Techniques in Cultural Heritage Materials Characterisation

Contribution of X-ray Fluorescence Techniques in Cultural Heritage Materials Characterisation

Editors

Anna Galli
Letizia Bonizzoni

MDPI • Basel • Beijing • Wuhan • Barcelona • Belgrade • Manchester • Tokyo • Cluj • Tianjin

Editors
Anna Galli
Dipartimento di Scienza dei
Materiali, Università degli
Studi Milano-Bicocca,
20125 Milano, Italy

Letizia Bonizzoni
Dipartimento di Fisica,
Università degli Studi di
Milano, 20122 Milano, Italy

Editorial Office
MDPI
St. Alban-Anlage 66
4052 Basel, Switzerland

This is a reprint of articles from the Special Issue published online in the open access journal *Applied Sciences* (ISSN 2076-3417) (available at: https://www.mdpi.com/journal/applsci/special_issues/Cultural_Heritage_Materials_Characterisation).

For citation purposes, cite each article independently as indicated on the article page online and as indicated below:

LastName, A.A.; LastName, B.B.; LastName, C.C. Article Title. *Journal Name* **Year**, *Volume Number*, Page Range.

ISBN 978-3-0365-4867-8 (Hbk)
ISBN 978-3-0365-4868-5 (PDF)

Contents

About the Editors vii

Preface to "Contribution of X-ray Fluorescence Techniques in Cultural Heritage Materials
Characterisation" ix

Anna Galli and Letizia Bonizzoni
Contribution of X-ray Fluorescence Techniques in Cultural Heritage Materials Characterization
Reprinted from: *Appl. Sci.* **2022**, *12*, 6309, doi:10.3390/app12136309 1

Giulia Ruschioni, Francesca Micheletti, Letizia Bonizzoni, Jacopo Orsilli and Anna Galli
FUXYA2020: A Low-Cost Homemade Portable EDXRF Spectrometer for Cultural Heritage
Applications
Reprinted from: *Appl. Sci.* **2022**, *12*, 1006, doi:10.3390/app12031006 5

Francesca Volpi, Giacomo Fiocco, Tommaso Rovetta, Claudia Invernizzi, Michela Albano,
Maurizio Licchelli and Marco Malagodi
New Insights on the Stradivari "Coristo" Mandolin: A Combined Non-Invasive Spectroscopic
Approach
Reprinted from: *Appl. Sci.* **2021**, *11*, 11626, doi:10.3390/app112411626 21

Cristina Fornacelli, Vanessa Volpi, Elisabetta Ponta, Luisa Russo, Arianna Briano,
Alessandro Donati, Marco Giamello and Giovanna Bianchi
Grouping Ceramic Variability with pXRF for Pottery Trade and Trends in Early Medieval
Southern Tuscany. Preliminary Results from the Vetricella Case Study (Grosseto, Italy)
Reprinted from: *Appl. Sci.* **2021**, *11*, 11859, doi:10.3390/app112411859 33

Ana Fragata, Jorge Ribeiro, Carla Candeias, Ana Velosa and Fernando Rocha
Archaeological and Chemical Investigation on the High Imperial Mosaic Floor Mortars of the
Domus Integrated in the Museum of Archaeology D. Diogo de Sousa, Braga, Portugal
Reprinted from: *Appl. Sci.* **2021**, *11*, 8267, doi:10.3390/app11178267 55

Anna Mazzinghi, Chiara Ruberto, Lisa Castelli, Caroline Czelusniak, Lorenzo Giuntini, Pier
Andrea Mandò and Francesco Taccetti
MA-XRF for the Characterisation of the Painting Materials and Technique of the *Entombment of
Christ* by Rogier van der Weyden
Reprinted from: *Appl. Sci.* **2021**, *11*, 6151, doi:10.3390/app11136151 69

Leandro Sottili, Laura Guidorzi, Anna Mazzinghi, Chiara Ruberto, Lisa Castelli, Caroline
Czelusniak, Lorenzo Giuntini, Mirko Massi, Francesco Taccetti, Marco Nervo, Stefania De
Blasi, Rodrigo Torres, Francesco Arneodo, Alessandro Re and Alessandro Lo Giudice
The Importance of Being Versatile: INFN-CHNet MA-XRF Scanner on Furniture at the CCR "La
Venaria Reale"
Reprinted from: *Appl. Sci.* **2021**, *11*, 1197, doi:10.3390/app11031197 83

Jacopo Orsilli, Anna Galli, Letizia Bonizzoni and Michele Caccia
More than XRF Mapping: STEAM (Statistically Tailored Elemental Angle Mapper) a Pioneering
Analysis Protocol for Pigment Studies [†]
Reprinted from: *Appl. Sci.* **2021**, *11*, 1446, doi:10.3390/app11041446 97

Alessandra Gianoncelli, George Kourousias, Sebastian Schoeder, Antonella Santostefano,
Maëva L'Héronde, Germana Barone, Paolo Mazzoleni and Simona Raneri
Synchrotron X-ray Microprobes: An Application on Ancient Ceramics
Reprinted from: *Appl. Sci.* **2021**, *11*, 8052, doi:10.3390/app11178052 121

About the Editors

Anna Galli

Anna Galli, Associate Professor of Applied Physics at the Department of Materials Science, University of Milano-Bicocca. Graduated in Physics, specialized in Materials Science, her research topics focuses on radiation physics linked to dosimetry and spectroscopy applied to Cultural Heritage. Scientific coordinator of LAMBDA (Laboratory of Milano Bicocca University for Dating and Archeometry) at Department of Material Science of University of Milano-Bicocca. Since 2020 until present, Member of scientific board of Associazione Italiana di Archeometria (AIAr).

Letizia Bonizzoni

Letizia Bonizzoni, masters degree in Physics (1998) and PhD in Analytical Chemistry (2006) at Milan University. Her research mainly deals with non-destructive analyses of cultural heritage; her main interest is devoted to the X-Ray fluorescence technique, mostly concerning the development of new methods of analysis on archaeometric materials. Applied research has been made on metallic alloys (coins and little artistic objects), glasses and enamels, ceramics, and pigments (painting, frescoes and illuminated manuscripts). She has collaborated with Italian Regional Boards for Cultural Heritage and Museums and co-operated with other research groups in Italy and worldwide.

Preface to "Contribution of X-ray Fluorescence Techniques in Cultural Heritage Materials Characterisation"

The editors would like to thank all the authors who contributed to this Special Issue, all the peer reviewers for the time spent to improve the manuscripts and the editorial team of Applied Sciences for their work. We are deeply in debt to Daria Shi for her suggestions and support.

This Special Issue would not have been surfaced without the passion transmitted to both editors by Mario Milazzo, pioneer of archaeometry in Italy and great expert of XRF.

Anna Galli and Letizia Bonizzoni
Editors

applied sciences

Editorial

Contribution of X-ray Fluorescence Techniques in Cultural Heritage Materials Characterization

Anna Galli [1],* and Letizia Bonizzoni [2],*

[1] Department of Materials Science, University of Milano-Bicocca, Via R. Cozzi 55, 20125 Milan, Italy
[2] Department of Physics, University of Milan, Via Celoria 16, 20133 Milan, Italy
* Correspondence: anna.galli@unimib.it (A.G.); letizia.bonizzoni@mi.infn.it (L.B.)

Citation: Galli, A.; Bonizzoni, L. Contribution of X-ray Fluorescence Techniques in Cultural Heritage Materials Characterization. *Appl. Sci.* **2022**, *12*, 6309. https://doi.org/10.3390/app12136309

Received: 9 June 2022
Accepted: 20 June 2022
Published: 21 June 2022

Publisher's Note: MDPI stays neutral with regard to jurisdictional claims in published maps and institutional affiliations.

1. Introduction

Archaeometry and cultural heritage have lately taken great advantage of developments in scientific techniques, offering valuable information to archaeology, art history, and conservation science, involving both instrumental and non-instrumental approaches. Among the possible techniques, X-ray Fluorescence (XRF) has become one of the most applied techniques for cultural heritage elemental material characterization, due to its user-friendliness; fast, short acquisition times; portability; and most of all, its absolutely non-destructive nature. For this reason, besides often being a first choice for a preliminary overall materials investigation, XRF spectrometers and spectra data handling methods have continuously improved, giving rise to many variations of the same technique; portable spectrometers, micro-probes, and large area scanners are all variations of a very flexible technique.

2. Review of Issue Contents

This Special Issue collects papers dealing with several analytical techniques and applications related to XRF spectroscopy, with special attention toward the application to cultural heritage materials. Paper subjects include instrumentation and technical developments, case studies on various materials requiring methodological innovations, and new data handling. Beside traditional XRF configuration, both energy dispersive and wave dispersive, Macro-XRF and the use of synchrotron radiation have also been considered. Indeed, most of the variations of XRF spectroscopy have been considered: portable or mapping spectrometers, synchrotron based XRF, and synergic association with other non-destructive analytical techniques.

In the paper by Ruschioni and co-authors [1], the project for a low-cost home-made XRF portable spectrometer is presented, for all those cases where there is not enough financial support to buy a commercial device. The FUXYA2020 spectrometer was intended mostly for cultural heritage (CH) applications and meets the requirements for both low Z matrix objects, such as glasses and ceramics, and medium-high Z materials, such as metals. First application shown (on pigments, ceramics and gold alloys) demonstrate that a simple and inexpensive prototype can be of great help for a rapid and reliable characterization of cultural heritage materials whenever commercial devices are unaffordable.

Portable instrumentation was also used in [2,3]. In the former paper, a mandolin by Antonio Stradivari has been investigated for the first time by non-invasive reflection Fourier transformed infrared (FT-IR) spectroscopy and X-ray fluorescence (XRF) on different areas previously selected by UV-induced fluorescence imaging. The combined spectroscopic approach allowed us to hypothesize original materials and finishing procedures similar to those used in violin making, and XRF results proved to be essential to support FT-IR findings and to detect possible iron-based pigments in the finishing layers. The paper by Fornacelli et al [3] presents a provenance study on ceramic sherds by means of portable XRF instrumentation coupled with statistical analysis. Indeed, the combined application of this kind of analytical technique and statistical analysis to the investigation of a large repertory of ceramic fragments allowed for the grouping of the assemblage by identifying geochemical clusters. Assuming a correlation between compositional patterns and local production centers, cultural relations between communities and regions were investigated starting from the pottery manufacturing and trade.

WD-XRF (wave dispersive XRF) was instead exploited in the work by Fragata and co-authors [4], where the floor mortar layers of the high imperial mosaics of a Roman housing in Braga, Portugal, were investigated, showing that their composition was clearly related to the stratigraphic position and to the external conditions and treatments to which they were submitted.

A few papers were devoted to macro-XRF (MA-XRF) instrumentation and related issues. The traditional field of application in CH materials, i.e., painting, was the subject of a research in the frame of the INFN-CHNet, the Cultural Heritage Network of the Italian National Institute of Nuclear Physics. They designed and developed an MA-XRF scanner easy to transport for in situ use. In the paper submitted for this special issue [5], they present results of a painting by the Flemish artist Rogier van der Weyden, belonging to the collection of the Uffizi gallery in Florence, Italy. The painting was analyzed during conservation treatments at the Opificio delle Pietre Dure in Florence. Also in this case, MA-XRF proved to be a powerful technique that can be easily utilized as an early non-invasive and non-destructive analytical method as a guide for a subsequent, more accurate, scientific analysis. The same scanner was used in the work by Sottili et al. [6] for an uncommon application to furniture at the Centro di Conservazione e Restauro "La Venaria Reale" (Turin, Italy), a leading conservation center in the field. The use of the MA-XRF technique on furniture has provided information on the elemental–spatial distribution of the decorative layers, such as gilding and ivory, on the polychromatic surfaces of a chinoiserie lacquered cabinet of the 18th century and a desk by Pietro Piffetti (1701–1777).

Again on MA-XRF, the work by Orsilli and co-authors [7] summarizes the advantages and limitations of MA-XRF, here considered as an imaging technique. This allows a better synergy with other hyperspectral methods, or the combination with spot investigations. A pioneering analysis protocol (STEAM) based on the spectral angle mapper algorithm is also presented, unifying the MA-XRF standard approach with punctual XRF, exploiting information from the mapped area as a database to extend the comprehension to data outside the scanned region, and working independently from the acquisition set-up.

Last, but of great interest, is a paper regarding synchrotron X-ray μ- and nano-probes applied to CH materials, in particular to ancient ceramics study [8]. Fine and varnished wares are the case study for exploring challenges offered by synchrotron X-ray microprobes optimized to collect microchemical and phase-distribution maps, capable of providing relevant clues for discriminating workshops and exploring technological aspects, which are fundamental in answering the current archaeological questions on ceramic findings.

Author Contributions: Conceptualization, A.G. and L.B.; writing—original draft preparation, L.B.; writing—review and editing, A.G.; supervision, A.G. and L.B. All authors have read and agreed to the published version of the manuscript.

Funding: This research received no external funding.

Acknowledgments: The editors would like to thank all the authors who contributed to this Special Issue, all the peer reviewers for the time spent to improve the manuscripts and the editorial team of Applied Sciences for their work. We are deeply in debt with editor for her suggestions and support. This Special Issue is dedicated to the loving memory of Prof. Mario Milazzo, a pioneer of Archaeometry in Italy, awarded in 2002 the Gold Medal for Culture by the Italian President. He is remembered as a generous and pleasant man with an insightful, logical mind, who was able to find an appropriate joke for every situation. Many of us, following his footsteps in the research field of applied physics for Cultural Heritage, still appreciate his vision, teaching and impact on our lives.

Conflicts of Interest: The authors declare no conflict of interest.

References

1. Ruschioni, G.; Micheletti, F.; Bonizzoni, L.; Orsilli, J.; Galli, A. FUXYA2020: A Low-Cost Homemade Portable EDXRF Spectrometer for Cultural Heritage Applications. *Appl. Sci.* **2022**, *12*, 1006. [CrossRef]
2. Volpi, F.; Fiocco, G.; Rovetta, T.; Invernizzi, C.; Albano, M.; Licchelli, M.; Malagodi, M. New Insights on the Stradivari Coristo Mandolin: A Combined Non-Invasive Spectroscopic Approach. *Appl. Sci.* **2021**, *11*, 11626. [CrossRef]
3. Fornacelli, C.; Volpi, V.; Ponta, E.; Russo, L.; Briano, A.; Donati, A.; Giamello, M.; Bianchi, G. Grouping Ceramic Variability with pXRF for Pottery Trade and Trends in Early Medieval Southern Tuscany. Preliminary Results from the Vetricella Case Study (Grosseto, Italy). *Appl. Sci.* **2021**, *11*, 11859. [CrossRef]
4. Fragata, A.; Ribeiro, J.; Candeias, C.; Velosa, A.; Rocha, F. Archaeological and Chemical Investigation on the High Imperial Mosaic Floor Mortars of the Domus Integrated in the Museum of Archaeology D. Diogo de Sousa, Braga, Portugal. *Appl. Sci.* **2021**, *11*, 8267. [CrossRef]
5. Mazzinghi, A.; Ruberto, C.; Castelli, L.; Czelusniak, C.; Giuntini, L.; Mandò, P.; Taccetti, F. MA-XRF for the Characterisation of the Painting Materials and Technique of the Entombment of Christ by Rogier van der Weyden. *Appl. Sci.* **2021**, *11*, 6151. [CrossRef]
6. Sottili, L.; Guidorzi, L.; Mazzinghi, A.; Ruberto, C.; Castelli, L.; Czelusniak, C.; Giuntini, L.; Massi, M.; Taccetti, F.; Nervo, M.; et al. The Importance of Being Versatile: INFN-CHNet MA-XRF Scanner on Furniture at the CCR "La Venaria Reale". *Appl. Sci.* **2021**, *11*, 1197. [CrossRef]
7. Orsilli, J.; Galli, A.; Bonizzoni, L.; Caccia, M. More than XRF Mapping: STEAM (Statistically Tailored Elemental Angle Mapper) a Pioneering Analysis Protocol for Pigment Studies. *Appl. Sci.* **2021**, *11*, 1446. [CrossRef]
8. Gianoncelli, A.; Kourousias, G.; Schöder, S.; Santostefano, A.; L'Héronde, M.; Barone, G.; Mazzoleni, P.; Raneri, S. Synchrotron X-ray Microprobes: An Application on Ancient Ceramics. *Appl. Sci.* **2021**, *11*, 8052. [CrossRef]

Article

FUXYA2020: A Low-Cost Homemade Portable EDXRF Spectrometer for Cultural Heritage Applications

Giulia Ruschioni [1], Francesca Micheletti [1], Letizia Bonizzoni [1,*], Jacopo Orsilli [2] and Anna Galli [2]

[1] Department of Physics, University of Milan, Via Celoria 16, 20133 Milan, Italy; giulia.ruschioni@studenti.unimi.it (G.R.); francesca.micheletti1@studenti.unimi.it (F.M.)
[2] Department of Materials Science, University of Milano-Bicocca, Via R. Cozzi 55, 20125 Milan, Italy; j.orsilli@campus.unimib.it (J.O.); anna.galli@unimib.it (A.G.)
* Correspondence: letizia.bonizzoni@mi.infn.it

Abstract: The project FUXYA2020 was intended to design and prototype a low-cost basic energy dispersive X-ray fluorescence spectrometer for all those cases where there is not enough financial support to buy a commercial device. Indeed, home-made instruments are ideal when funds are low but constant over the years, as this approach allows the costs to be spread over a longer period of time. The FUXYA2020 was intended mostly for cultural heritage (CH) applications: we optimized the geometry to meet the requirements for both low Z matrix objects, such as glasses and ceramics, and medium-high Z materials, such as metals; besides, we designed a positioning system through Arduino components to obtain good results and repeatability for samples with a complex geometry. The FUXYA2020's performance was tested both for qualitative and quantitative analyses, the former on pigment layers, and the latter on gold-based certified alloys, exploiting Axil-QXAS software for data elaboration. The classification of ancient ceramics based on multivariate analysis obtained through R environment was also carried out. The qualitative data on pigments have also been compared with the same data obtained by a commercial XRF spectrometer, demonstrating how our very simple and inexpensive prototype can be of great help for a rapid and reliable characterization of cultural heritage materials whenever commercial devices are unaffordable.

Keywords: EDXRF; cultural heritage materials; pigment analysis; ceramics classification; metal alloy quantitative analysis

Citation: Ruschioni, G.; Micheletti, F.; Bonizzoni, L.; Orsilli, J.; Galli, A. FUXYA2020: A Low-Cost Homemade Portable EDXRF Spectrometer for Cultural Heritage Applications. *Appl. Sci.* **2022**, *12*, 1006. https://doi.org/10.3390/app12031006

Academic Editors: José A. González-Pérez and Frank Walther

Received: 21 December 2021
Accepted: 17 January 2022
Published: 19 January 2022

Publisher's Note: MDPI stays neutral with regard to jurisdictional claims in published maps and institutional affiliations.

1. Introduction

Energy dispersive X-ray fluorescence (EDXRF) is a very suitable analysis method for the examination of cultural heritage materials and is widely used, thanks to its relative straightforwardness of use and its totally non-destructive character, allowing for analysis without any sample preparation. EDXRF has numerous applications in this field, as it is able to detect elements from sodium to uranium, and in certain cases can go down to carbon [1], moreover the data treatment (e.g., qualitative analysis, PCA [2,3], semi-quantitative analysis [4–6], quantitative analysis) makes this technique suitable to answer numerous questions. Thus, EDXRF is commonly used to perform provenance studies on ceramics, coins and glass [3,7–9], to evaluate corrosion processes on metals [10,11], find restorations in paintings [12–14] or fake objects [15], or to identify worn out pigmentation in ancient polychromies [16–19]. These are just few examples of potential applications.

Anyway, we also must consider that XRF investigates deep into the artifacts (depending on the energy of the fluorescence radiation and the composition of the sample, the volume investigated can be a few tens of micrometers for light matrices to less than one micrometer for metals). The technique is then not always suitable when the aim is to distinguish the composition of the different layers composing the sample, and it is only a comprehensive elemental composition that is obtained. The only way to distinguish the layers is to get a priori knowledge of the composition of the sample, or to analyze the bulk if feasible. Nevertheless, new set-ups are now also being developed to analyze layered samples [20–22]: for example, performing an angular scanning [23–25], evaluating the ratio of the fluorescence lines [26,27], exploiting MA-XRF mapping [28–30], or through Monte Carlo simulations [11,31,32].

In the last twenty years, several portable and handheld EDXRF spectrometers have become commercially available, each with its own distinguishing features and fields of application. Their diversity of designs and their features affect the final price of this kind of instrumentation, making it accessible to a small circle of researchers. It must be pointed out that some hand-held instruments do not offer the easy possibility of working on raw spectra, but only to gain a quantitative evaluation of element concentrations [2,33]. This aspect makes these instrumentations fake user-friendly, as they require high skills to handle and customize the routines and analytical modes provided by the manufacturer. Moreover, not all the materials pertaining to the cultural heritage field can be described in terms of element concentration, as, for instance, is the case of pigments [34], where dilution, binder and layer thickness highly affect the possibility (and the meaningfulness) of having a quantitative evaluation. The same applies to the study of any layered sample, such as lustered or glazed ceramics [35], or to ink decorations [16]. From these starting considerations, in the last decade some research groups have developed their own homemade spectrometers; some of them clearly demonstrated that basic spectrometers can be useful for the characterization of CH materials held in the Museums [36]. In 2010, a research group from the Physics Institute of the UNAM, Mexico, developed SANDRA (sistema de analisis no destructivo por rayos X or system for non-destructive analysis using X-rays) especially for the study of Mexican Cultural Heritage collections [37]. This system is equipped with changeable tubes (namely, 75 W Mo, Rh and W X-ray tubes) and detectors (Amptek Si-PIN and Cd-Te detector) that are selected and combined depending on the elemental range of interest. One year later, LABEC laboratory of Istituto Nazionale di Fisica Nucleare (INFN), in collaboration with ICVBC of Consiglio Nazionale delle Ricerche (CNR), developed a custom-realized portable XRF spectrometer [38]. The system features a high efficiency for a wide range of elements down to sodium, thanks to the use of two tubes with different anodes and to the presence of a continuous helium flow in front of the tubes and the detector, with the declared aim to overcome the usually limited performance of compact XRF systems for in situ analyses. An interesting project was presented in Campos et al. [39], where a low-cost portable system for in situ elemental mapping by XRF was developed. The basic components (i.e., X-ray tube and detector) are very close to ours, but the final aim, and thus the design, are highly different, because, for a good scanning system, more attention must be given to the positioning system, the acquisition time and the ability to reconstruct elemental maps with good spatial resolution, while quantitative analyses is not considered.

Some homemade micro-XRF spectrometers were also studied [40], but their discussion is out of the scope of the present paper.

The aim of the FUXYA2020 (fluorescence X-ray analysis) project was to create a low-cost portable prototype for EDXRF analysis that was compact, lightweight, easily transportable, and attachable to a tripod for convenient positioning. In the FUXYA system, the two most expensive elements are obviously represented by the X-ray source and the detection unit. These are mandatory for a reliable spectrometer, but with a homemade project it is possible to spread the purchasing on different periods. In the following, we will discuss the design, the choice of elements and the performances of FUXYA2020 when it is used for the characterization of cultural heritage materials, namely pigments, ceramics, and metal alloys. Despite its basic features, the FUXYA spectrometer demonstrates its capability for work with rapid and reliable characterization on cultural heritage materials whenever commercial devices are unaffordable.

2. FUXYA2020 Design and Characteristics

2.1. X-ray Source

The X-ray source used is the MINI-X2 X-ray tube (Amptek, Bedford, MA, USA) with a maximum power of 4 W (50 kV, 200 μA) and a transmission rhodium anode. The choice of this material for the anode was made for several reasons: the characteristic lines of the K series of rhodium are present at relatively high energies and produce very definite Compton and Rayleigh diffusions that can be useful for the quantitative analysis of light matrix materials [3,6,41,42]; rhodium is rarely present in cultural heritage materials and this decreases the possibility of interference with the sample. Indeed, when choosing the anode, it is of the utmost importance to consider that the element of the anode cannot be evaluated in the samples, and that the anode line may interfere with the fluorescence line of the element of interest present in the sample. Employing a silver anode in the field of cultural heritage, for example, will limit the analysis of silver in metal sample, and thus, will limit the analytical capability of the instrument. In Table 1, we summarized the interferences of the four anode materials available for the MINI-X2 X-ray tube, where the inference is highlighted if the energy of anode characteristic line differs from the line of the element for less than 250 eV. The lines considered are Kα, Kβ, Lα, Lβ, Mα, while we do not consider the noise due to the detection system.

Table 1. Interference of the anode material lines on different elements, the interference is detected if any line of the analyte has an energy difference of less than 250 eV with any line of the anode.

Anode	L/K	K/K or L/L	L/M	M/K
Mo	P, S, Cl	Zr, Nb, Tc, Ru, Pt	Au, Hg, Tl, Pb, Bi, Po, At	
Rh	S, Cl, Ar	Tc, Ru, Pd, Ag	Po, At, Rn, Fr, Ra, Ac, Th, Pa	
Ag	Cl, Ar, K	Rh, Pd, Cd, In, Sn	Fr, Ra, Ac, Th, Pa, U	
W	Ni, Zn, Ge, Br, Kr, Rb, Sr, Y	Tm, Yb, Lu, Hf, Ta, Re, Os, Ir, Au, Hg		Al, Si, P
Au	Zn, Ge, Sr, Y, Zr, Nb, Mo	W, Os, Ir, Pt, Hg, Tl, Pb, At, Rn		P, S

The Mini-X2 focuses radiation in the designated output direction, and we verified the radiation levels external to the X-ray tube housing with the brass safety plug on. Nonetheless, the FUXYA system was housed in a radiation shielded chamber allowing the operator to work from outside the chamber. For maximum safety, the door switch and the red warning light were integrated with the Mini-X2 interlock circuit.

Two brass collimators with aluminum inserts (with 1 mm and 2 mm diameter holes) are provided to geometrically decrease the diameter of the outgoing X-ray beam in order to have a fairly small measuring spot, suitable for applications in cultural heritage [43].

The Mini-X2 is also equipped with a set of filters to modify the output spectrum of the tube, filters are used to improve the signal-to-noise ratio for the elements of interest and attenuate the unwanted characteristic lines of the anode and the materials of which the X-ray tube is made [44]. In Figure 1, we report the different spectra produced by the X-ray tube when equipped with the available filters, namely: Al (254 μm), Cu, Mo, Ag, W (25.4 μm). Spectrum emitted with no filter was given by the producer, while filtered ones were obtained by calculating for each energy the radiation absorption given by the specific filter. The choice of filter usually depends on the source anode. If the K lines of the anode are used to excite the sample, one can employ filters made of the same material of the anode, absorbing the energies above the absorption edge, and obtaining a quasi-bichromatic source. Otherwise, we can employ the element with atomic number $Z = Z_A - 2$ or $Z = Z_A - 1$, where Z_A is the atomic number of the anode; in this case the absorption edge of the filter falls between the Kα and Kβ lines of the source, highly absorbing the latter, and transmitting the former, obtaining in this way a quasi-monochromatic source. Finally, usually an Al filter can be employed to absorb the bremsstrahlung at low energies, increasing the signal to noise, S/N, ratio for low Z elements.

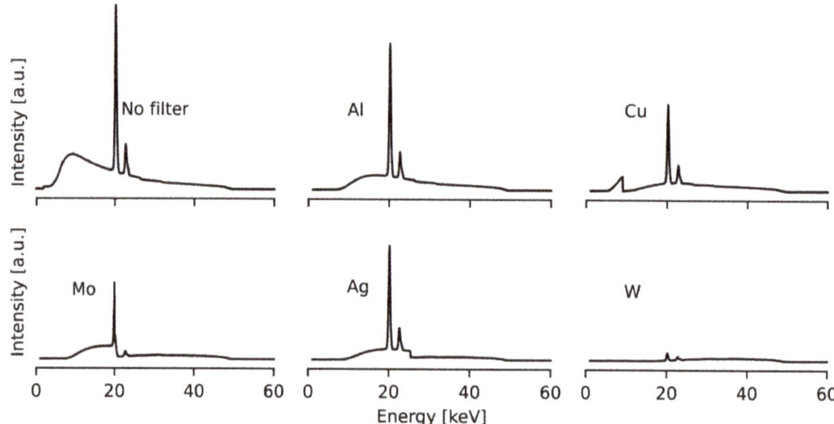

Figure 1. Attenuation of X-ray tube spectra by Al (254 µm), Cu, Mo, Ag and W (25.4 µm) filters.

2.2. Detector

The X-ray detector used is a complete SDD spectrometer (XGL-SPCM-DANTE-25 model) from XGLAB Bruker Nano Analytics, Milan, Italy, with 17 mm^2 active area, 500 µm thickness and 12.5 µm Be window. The energy resolution for an SDD detector is usually given as FWHM of the Mn Kα line (5.9 keV); for our spectrometer, the nominal energy resolution is 140 eV with peak-shaping time of 96 ns, or 125 eV with 1 µs peak-shaping time. In our working conditions (0.8 µs shaping time) we measured the detector resolution on a pure metal Mn target obtaining 130 eV as expected; for count rate of about 1 Mcps, we have a dead time of about 2%. In order to better detect the X-ray fluorescence emitted by the sample eliminating fluorescence from the detector itself, it is necessary to use collimators in front of the detector Be window, in addition to the internal one. We used a 2 mm lead collimator completely lined in pure aluminum (total thickness of the collimator 5 mm), so that no detectable lines are added. To allow an easy change of the collimator, whose best diameter can vary depending on the object to be studied, a cap for the detector window has been designed and 3D printed. This expedient also introduces an indirect protection to the thin Be window, but slightly increases the distance between the detector and the sample. In the geometric configuration of the FUXYA system, the angle between the X-ray tube and the detector was set at 90° (Figure 2) in order to better separate the Compton scattering peak from the Rayleigh scattering peak and to obtain information about light elements when necessary [45].

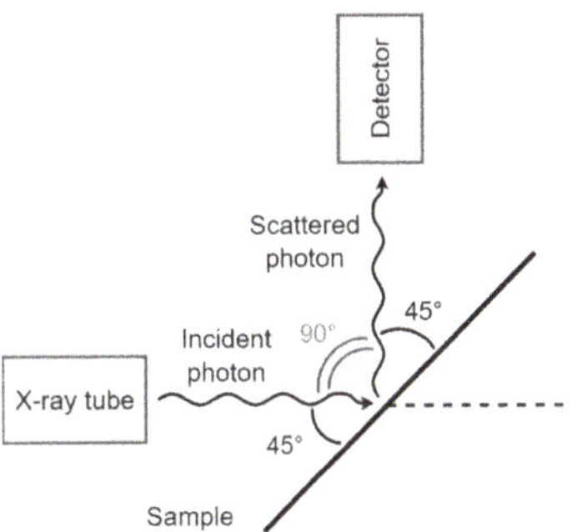

Figure 2. Geometrical configuration of FUXYA2020 spectrometer.

2.3. Assembling

Once the geometric configuration of the tube-detector was established, the final integration between these two components was designed using a 3D design software (Figure 3). The enclosure includes a base with four support feet at the bottom, and three holes to secure a Manfrotto 200PL rapid plate so that it can be attached to a tripod. In addition, the X-ray source and the detector unit have been positioned so that the intersection point between the X-ray beam from the X-ray tube and the detector axis, i.e., the "focus" of the instrument, is about 10 mm away from the base of the instrument; the distance between the X-ray tube and the sample is thus about 14 mm, while that between the sample and the detector is about 34 mm. The current distance between the instrument and the sample can be reduced simply by shifting the components on their axis; nevertheless, choosing a 45°–45° geometry, the size of the detector head must be considered. If a significantly lower distance is preferred, the geometry should be changed, either by reducing the scattering angle and keeping the irradiation angle equal to the detection angle, or by maintaining a 90° scattering angle with a different geometry (such as a 60°–30° geometry).

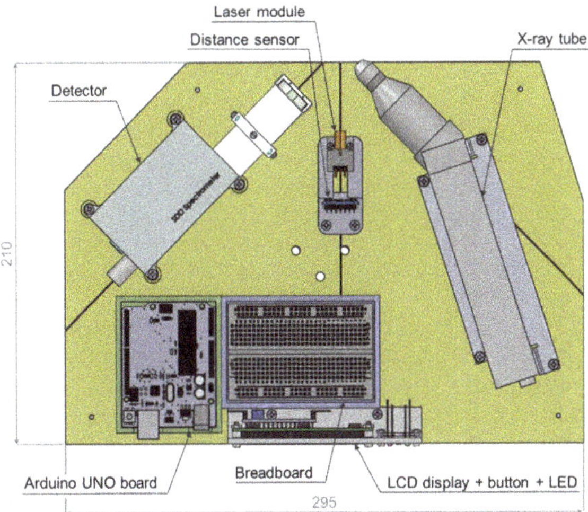

Figure 3. Positioning of the components for FUXYA2020 spectrometer. Dimensions are given in mm.

Thanks to the 3D design software [46], the measurement spot is easily calculated on a flat sample: when using the 1 mm collimator for the X-ray tube, the area irradiated on the sample is about 8 mm^2 and has an obviously elliptical shape (Figure 4). To complete the FUXYA2020 system, a cover was created, which includes a cooling system exploiting two USB fans (5 V), one positioned at the top, in correspondence with the detector, and the other laterally, close to the X-ray tube (Figure 5).

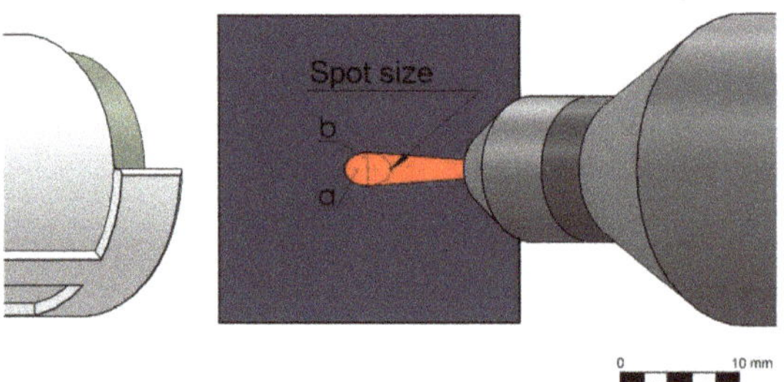

Figure 4. Sample irradiated area when using the 1 mm collimator for the X-ray tube (a $\cong$ 1.91 mm, b $\cong$ 1.35 mm).

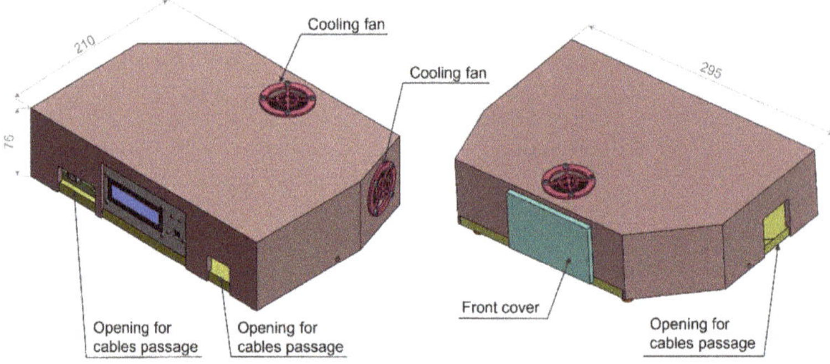

Figure 5. FUXYA2020 external case. Dimensions are given in mm.

Furthermore, inside the case we placed a positioning system run through an Arduino UNO board and composed of the following electronic components (Figure 3): a laser sensor module (KY-008) with a spot size of 5 mm (outer diameter of 6 mm) to verify the investigated area, and a VL6180X time of flight (ToF) distance sensor to position the sample in correspondence with the focus of the instrument, maximizing the spectra intensity and allowing repeatable and comparable measurements. The VL6180X is an optical module that includes a proximity sensor, an ambient light sensor, and a VCSEL (vertical cavity surface emitting laser) light source which emits in the near-IR at a wavelength of 850 nm. The distance sensor has a spatial resolution of 1 mm and an accuracy of about 1% at 50 mm distance. We did not choose an ultrasonic sensor as it is not suitable for non-flat surfaces; moreover, differently from an IR sensor (whose range of distance is also out of the scope of the present project), ToF measure should be independent from the sample reflectance. We also integrated an I2C 1602 LCD display to show the distance measured by the sensor, a button to easily turn on and off the laser module and two LED diodes that light up according to the distance from the sample. All these components can be bought at a low price, according to the financial availability of the research group.

To make a low-cost instrument, all the structural components of the FUXYA system were 3D printed in PLA (polylactic acid) except for the base, which was cut out from a PVC sheet for greater structural strength (Figure 6). The maximum footprint of the FUXYA system is 295 × 210 × 76 mm, similar to the size of an A4 sheet of paper.

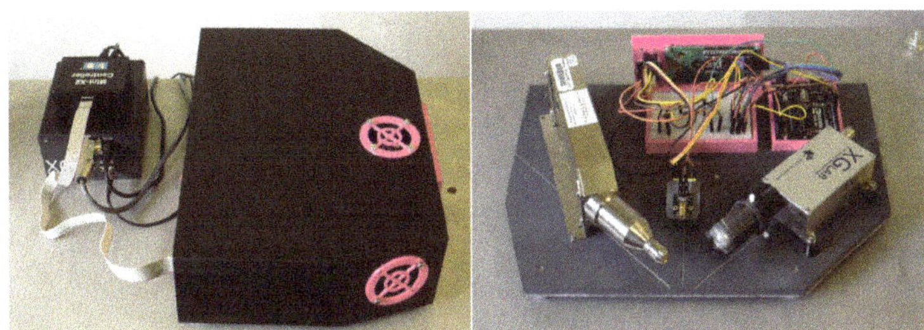

Figure 6. FUXYA spectrometer and its internal set up.

3. FUXYA2020 Performance on Real Cases in Cultural Heritage Field

The FUXYA's performance was tested on the qualitative analyses of pigment layers, on semi-quantitative analyses of archaeological ceramics for provenance classification and on quantitative analyses on gold alloys, thus covering three main materials in the CH field. For the first material, we had compared the minimum detection limit (MDL) for the pigment characterizing elements obtained from FUXYA and the one achieved with a commercial spectrometer. In the case of ceramics studies, the results obtained by applying a semiquantitative analysis were checked with the archaeological evidence and finally the element concentrations from quantitative analysis on metal alloys were compared with the certified values.

3.1. Pigments

In the last 20 years, EDXRF has proved to be a very suitable and widely used tool for the examination of paintings, also for its fast response and its totally non-invasive character [47]. Analysis of pigment layers using the FUXYA was conducted on laboratory-made panels (Figure 7), choosing single layers (monolayers) of historical and modern pigments, as reported in Table 2. The unfiltered tube spectrum was used as incident radiation, thanks to the low background of the instrument at low energies. The operating conditions were 40 kV and 0.06 mA, with an acquisition time of 200 s for each measurement. As many factors influence the pigment stratigraphy (binder, dilution, and thickness of the layers, which also depend on many factors) when dealing with pigment layers, it is more useful and sensible to perform a qualitative analysis, as a quantitative analysis is not only not feasible but will not be useful to characterize the nature of the pigment.

Figure 7. Analysis of pigment layers by FUXYA spectrometer: the laser beam for positioning and the red light indicating the correct distance are visible.

Table 2. Pigment layers considered and elements detected by FUXYA spectrometer. (*) Layer B6, labelled as Mount Amiata cinnabar, revealed to be a counterfeit pigment.

Layer	Pigment	Characterizing Elements from Known Composition	Detected Elements (Traces)
A1	Natural Lapis Lazuli	S, Si	(Si, K, Fe)
A2	Smalt Blue	Co, K, Si	Co, K, (Si, Fe)
A3	Cobalt Blue	Co	Co, Zn
A4	Azurite	Cu	Cu (Fe)
A5	Verdigris	Cu	Cu
A6	Green Earth	Fe, K	Fe, K (Mn, Ti)
A7	Raw Earth Umber	Fe, Mn	Fe, Mn (Ti)
A8	Tin-Lead Yellow	Pb, Sn	Pb, Sn, (Cu)
A9	Minium	Pb	Pb
A10	Cinnabar	S, Hg	Hg (S)
A11	Red Ochre	Fe	Fe (K, Ti, V)
A12	Burnt Sienna	Fe (Mn)	Fe (K, Ti, V)
B1	Cerulean Blue	Co, Sn	Co, Sn, Zn, Ti
B2	Naples Yellow	Pb, Sb	Pb, Sb, Zn (V, Mn)
B3	Malachite	Cu	Cu (Fe)
B5	Chromium Orange	Cr, Pb	Cr, Pb
B6	Mount Amiata Cinnabar *	S, Hg	Ti, Ba, Zn, Sr

Test panels of pure pigment layers were painted on different substrates, with different binders: panel A (binder: egg tempera) is a wooden board with gypsum preparation, while panel B (acrylic binder) is an industrial canvas applied on cardboard. The pigments used for these tests were natural and artificial lapis lazuli, smalt blue, cobalt blue, azurite, verdigris, titanium green, umber, lead yellow, minium, cinnabar, red ochre, burnt sienna (panel A) and cerulean blue, Naples yellow, malachite, cinnabar, chromium orange (panel B). As the pigments layers are very thin, during an XRF measurement is possible to detect elements coming from the preparation underneath (except in cases where heavy metals are used, such as lead), thus preparation of both panels was also investigated as reference. For panel, A, Ca, Sr and S traces were detected from gypsum preparation, with impurities of P, Fe, Cu, Zn, Zr; for panel B, the analyses determined the presence of titanium white (Ti dioxide) and Ca, with impurities of P, Fe, Cu, and Zn. These elements are reported in Table 2 only when attributable to pigment layers; each spectrum also presents the K emission of Ar from the atmosphere. As evident from Table 2, all pigments were positively detected within the XRF limits, except sample B6, Mount Amiata cinnabar, which revealed itself to be a case of forgery, as no presence of Hg (nor any other red chromophore) was detected. The barium presence can be attributed to barium sulphate, commonly used as a pigment extender with lakes. A rapid check with other analytical techniques confirmed the use of a red lake with a mixture of white pigments, coherently with FUXYA results. Further comments should also be done on sample A12, burnt sienna earth, where Mn could be expected. Indeed, sienna is a yellow ochre, containing from 50% to more than 70% iron oxide (Fe_2O_3), (the higher the iron concentration, the better the quality) [47] and, even if it is often considered a mixture of Mn and Fe oxides, Mn oxide (gray or black in color) is quite low, and it is present in quantities up to 10% [48], while a greater amount of Mn oxides are typical of umbers. Coherently, some spectra databases [49] indicate sienna earth also in the absence of Mn.

In Figure 8, the spectra acquired with FUXYA spectrometer on lapis lazuli layered pigment (sample A1 in Table 2) is reported; the Si Kα lines are evident in the enlargement of low energy region.

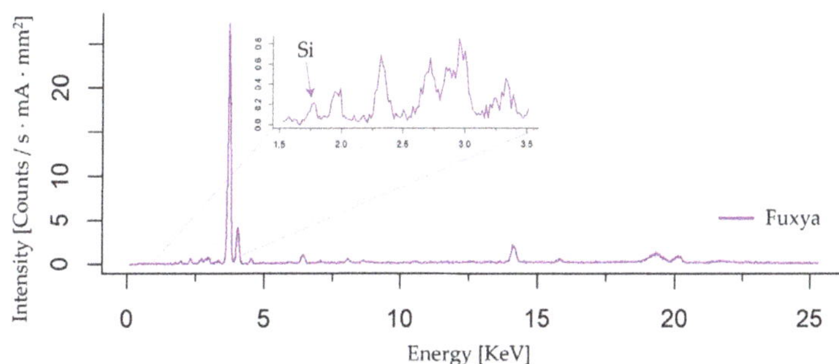

Figure 8. Spectra obtained by FUXYA for Layer A1, lapis lazuli (Si K line is highlighted in the blown up area).

Besides verifying the capability of our spectrometer to identify pigments, pigment spectra were also used to verify the ability to distinguish the signal with respect to fluctuations of the background. With this aim, peak areas were normalized to the acquisition time and anode current, and the instrumental limit of determination (ILD) was defined as: $ILD = 3\sqrt{cps(mA)}$, according to the definition usually given for the concentration MDL [50]. We did not normalize with respect to the irradiated surface on the sample, as we wanted to compare performance in standard working conditions for each spectrometer. We did not opt for the normally used MDL as, in this case, the passage to concentration values would be aleatory; anyway, ILD and MDL are proportional. Moreover, ILD depends on the instrument, the specimen matrix composition and the element considered and can thus be used for comparing the performance of different instruments used in the same analytical context. The results (see Table 3) were then compared with those obtained measuring the same samples with a commercial portable micro-XRF spectrometer (Bruker ARTAX 200) equipped with a Mo anode X-ray tube, collimated down to 0.65 mm in diameter (excited sample area of 0.33 mm^2). The Artax 200 mounts an SDD detector and exploits the principle of optical triangulation for the adjustment of the distance between sample and measuring head, and a CCD camera to visualize the sample. The system has an exchangeable filter slide with three filter positions; in this work, no filter has been used. The working conditions were 40 kV and 0.6 mA with an acquisition time of 200 s for each measurement; the distance between the analyzed spot and both the source and the detector was 1.3 cm. Bruker Spectra 5.1 software was used to perform peak deconvolution. As evident from Table 3, even if the results are similar to those obtained in a former publication [51], they cannot be directly compared, as in that case the Artax source was filtered by a 12.5 μm Mo transmission filter. It is worth noting that, in this same paper, the MDL for the Assing Lithos portable and Niton handheld spectrometers were also considered; in particular, our spectrometer can be compared with Lithos for dimensions, features and portability, showing a better performance when using non-filtered X-ray tube emission in standard working conditions.

When dealing with qualitative XRF analysis, an obvious limit is the difficulty in detecting low energy signals, as for instance for Sb and Sn L lines, and Si K lines, because the probability of emission is lower for light elements and for L lines. To better visualize the different performances, in Figure 9, we report the comparison between FUXYA and Artax spectra for samples A8 and B2 in their standard operating conditions. The spectra of the considered yellow layers are reported in linear scale for both FUXYA and Artax spectrometers. Spectra have been normalized for the measurement time, the applied amperage and the irradiated area. The FUXYA spectra show a lower intensity, but also a lower background; for this reason, we can see trace element peaks in a clearer way. Moreover, the better resolution allows us to better distinguish Si and P peaks. It is worth saying that even though Sn and Sb can be easily detected by their K lines with FUXYA device when non-filtered radiation is used, we are considering their L emissions as a clear example of the sensitivity at lower energies which is increased in the FUXYA system thanks to the lower background. Moreover, the detection of both K and L series for such elements can help in reconstructing the stratigraphy of the pigment layers

Table 3. Comparison of ILD in cps/mA of different elements.

Element	FUXYA	Artax	Element	FUXYA	Artax	Element	FUXYA	Artax
Si (K_α)	4.42	9.32	Mn (K_α)	16.18	14.32	Zr (K_α)	12.85	10.79
P (K_α)	5.68	9.50	Fe (K_α)	17.68	11.65	Sn (K_α)	11.68	7.87
S (K_α)	7.14	8.69	Co (K_α)	15.54	13.99	Sn (L_α)	6.54	7.89
K (K_α)	9.80	8.90	Ni (K_α)	9.87	6.52	Sb (K_α)	11.32	5.28
Ca (K_α)	9.57	10.28	Cu (K_α)	18.12	7.93	Sb (L_α)	8.96	10.08
Ti (K_α)	17.62	12.20	Zn (K_α)	17.54	11.45	Hg (L_α)	9.25	7.57
Cr (K_α)	10.43	8.75	Sr (K_α)	11.49	11.78	Pb (L_α)	13.42	9.59

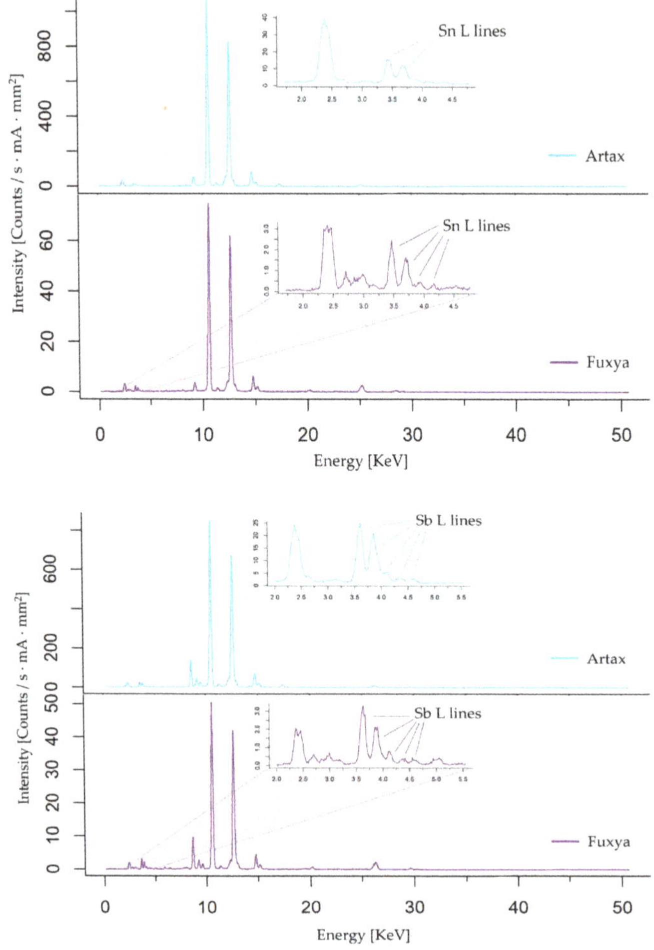

Figure 9. Spectra of the yellow layers A8 (**upper**) and B2 (**lower**) for both FUXYA and Artax spectrometers. Spectra have been normalized for the measurement time, the applied amperage, and the irradiated area.

Please note that the air path of the detection system is not the same: in fact, in FUXYA the detector is placed further away with respect to the Artax. This different geometry is responsible for the different order of magnitude in spectra intensities both for the different air paths, and for the subsequent different solid angles. In addition, the FUXYA spectra show the scattered L lined from Rh anode of the X-ray tube between 2.6 and 3.1 keV.

3.2. Ceramics

A semi-quantitative test on archaeological ceramics samples has also been done, verifying the possibility of using the FUXYA spectrometer for archaeological material classification. Eight Roman amphorae were considered, of which the factory, and thus the production place, were known from archaeological information, such as seals, as summarized in Table 4. No archaeological consideration will be made in this work, as this research is still in progress and we are now interested only in the FUXYA's performance, but it is worth noting that we chose a small group with the same local provenance, and a few samples of various non-local provenance to have an easy test for our set up.

Table 4. Analysed ceramics.

Sample	N. of Spectra	Archaeological Classification
4	3	Non-local (NA)
5	3	Non-local (MA)
7	3	Non-local (AA)
9	3	Non-local (BA)
14	2	local
15	2	local
16	2	local
17	4	local

We used the unfiltered continuous radiation from the X-ray tube and semi-quantitative evaluation, avoiding the determination of element concentrations [6].

The XRF spectra were acquired on fractures and/or cleaned areas; where necessary, cleaning was performed with pure water on some cotton wool to eliminate visible earth traces. A variable number of measuring points was considered for each sample to check the homogeneity of the artifacts [52]; for a few samples, Cr (7_3), Cu (7_1, 17_4) or Rb (9_3) showed peaks below the detection limit (see Section 3.1); in these few cases, to perform statistical analyses, we substituted the missing data with a random value between 0 and the detection limit itself [8,53]. All the 22 spectra were then considered for statistical analysis on closed data. Statistical elaboration was obtained by using R 3.6.3 [54], a free environment for statistical computing and graphics. It compiles and runs on UNIX platforms, Windows and MacOS. In particular, the FactoMineR 2.4 [55] and pca3d 0.10.2 [56] packages have been used. Data elaboration was performed once again in the perspective of a low-cost project, requiring no expensive licenses. PCA was applied, exploiting the correlation matrix as relative concentrations have different scales and we wanted to standardize data; relative concentration was closed to 100% for multivariate analyses and classification to eliminate the differences among samples due to different silicate presence or firing temperatures, which could induce a varying weight loss also in samples with similar raw materials. This procedure is particularly advisable whenever samples contain indefinite amounts of extraneous material, such as temper (i.e., crushed shell or crushed stone) [8,57].

The working conditions were 40 kV and 0.06 mA with an acquisition time of 200 s for each measurement. The detector proprietary software was used to perform peak deconvolution and calculate areas.

In Figure 10, the loading plot for the first two PCs is reported: even if the cumulative variance is relatively low (61.2%) and the samples are few, the various groups are well separated and allow the discrimination between different provenance which occupies different areas in the PCs cartesian plane. In fact, the local samples (14, 15, 16, 17) were grouped together, well-spaced from the non-local samples.

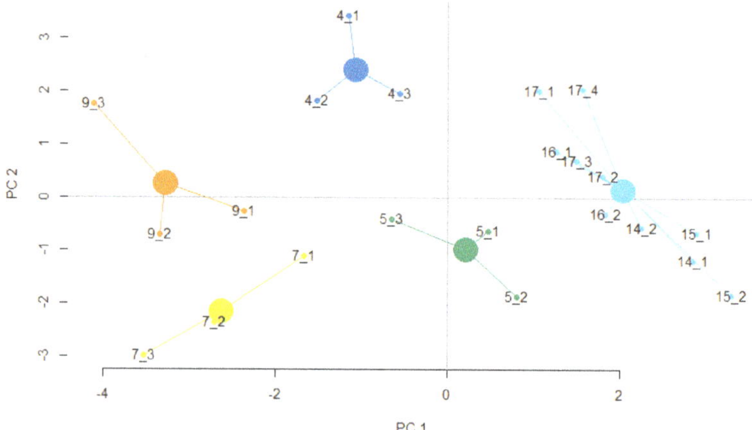

Figure 10. Loading plot for the first two PCs: cumulative variance is 61.2%; larger dots are the centroids for provenance groups.

3.3. Metals

The FUXYA's capability to get reliable quantitative evaluation on metal alloys was tested measuring the binary and ternary gold alloys (1 mm thickness) reported in Table 5 with the obtained results. We coupled the X-ray tube with a Cu filter (25.4 µm of thickness) so that incident radiation could be considered quasi-monochromatic at the Rh Kα line at 20.2 keV (see Section 2.1). Spectra deconvolution and elemental determination were made by AXIL/QXAS, a free quantitative analysis package developed by IAEA [58]. Calculations were made considering monochromatic excitation, single element reference standards (i.e., a pure Cu certified metal plate, a pure Ag certified metal plate, a pure Au certified metal plate), individual instrumental constants and applying normalization for the obtained concentrations. The acquisition parameters were 40 kV and 60 µA for 600 s. Characteristic emissions considered were Kα for Cu, and Lα for Ag and Au. The same was done for 100 s acquisitions, still obtaining the right concentrations, even if with higher errors, especially on Ag concentration (data not shown).

Table 5. Results obtained for quantitative analysis on certified gold alloys for monochromatic incident radiation. Errors are calculated by AXIL/QXAS considering statistical errors on both peaks and background fitting.

Alloy	Ag		Au		Cu	
	Certified	Calculated	Certified	Calculated	Certified	Calculated
AgAuCu1	25	25.07 ± 2.30	60	60.21 ± 0.22	15	14.72 ± 0.19
AgAuCu3	10	8.96 ± 1.45	80	80.79 ± 0.52	10	10.25 ± 0.17
AgAuCu5	-	-	80	79.52 ± 0.53	20	20.48 ± 0.12

4. Conclusions

The FUXYA project produced a compact and light prototype for a low-cost EDXRF portable spectrometer. The home-made solution allowed the costs of the X-ray tube and detector to be spread over a relatively long period of time; all the other components were cheap, and the software used for data analysis was free. The case and the structural components were 3D printed. The maximum dimensions were $295 \times 210 \times 76$ mm^3.

FUXYA2020 performance was tested both for qualitative and quantitative analyses, the former on pigments layers, and the latter on gold-based certified alloys, exploiting Axil-QXAS software for data elaboration. The classification of ancient ceramics based on multivariate analysis obtained through R environment was also carried out. Qualitative data on pigments have also been compared with the same data obtained by a commercial XRF spectrometer, demonstrating how our very simple and inexpensive device can be of great help for a rapid and reliable characterization of cultural heritage materials.

For the future, several improvements to the FUXYA system are foreseen, such as the use of a small camera that, together with the laser pointer, would create a better positioning system, based on the principle of optical triangulation, as in the Artax 200 spectrometer. In this way the distance sensor, not very reliable on reflective surfaces, could be replaced.

Another improvement is to use a Raspberry Pi instead of the Arduino UNO board. The main difference between these two systems is that Arduino UNO is a simple microcontroller board that is programmed using the Arduino IDE and can only run the compiled code; while Raspberry Pi is a single-board computer that can operate as a stand-alone system. The necessary software (Mini-X2 Controller for the X-ray tube and DANTE for the detector) could be installed directly on the Raspberry Pi, optimizing the communication between the programs. Eliminating Arduino and the breadboard would also save space so that the source controller and the detector electronics could be housed inside the case, achieving a fully integrated system. All this would lead to completely discarding the laptop in favor of an even more portable and compact system, with no cost increase.

Author Contributions: Conceptualization, L.B.; software, G.R.; validation, G.R., F.M. and J.O.; formal analysis, F.M. and G.R.; investigation, G.R., F.M. and J.O.; writing—original draft preparation, L.B. and G.R.; writing—review and editing, L.B. and A.G.; supervision, L.B. and A.G. All authors have read and agreed to the published version of the manuscript.

Funding: This research received no external funding.

Data Availability Statement: The data presented in this study are available on request from the corresponding author.

Acknowledgments: The authors would like to thank the Physics Department Mechanical Workshop of Università degli Studi di Milano for their suggestions and kind availability.

Conflicts of Interest: The authors declare no conflict of interest.

References

1. Uhlir, K.; Griesser, M.; Buzanich, G.; Wobrauschek, P.; Streli, C.; Wegrzynek, D.; Markowicz, A.; Chinea-Cano, E. Applications of a New Portable (Micro) XRF Instrument Having Low-Z Elements Determination Capability in the Field of Works of Art. *X-ray Spectrom.* **2008**, *37*, 450–457. [CrossRef]
2. Idjouadiene, L.; Mostefaoui, T.A.; Djermoune, H.; Ziat, F.; Bonizzoni, L. XRF Analysis of Ancient Numidian Coins: A Comparison between Different Kingdoms. *Eur. Phys. J. Plus* **2021**, *136*, 512. [CrossRef]
3. Idjouadiene, L.; Mostefaoui, T.A.; Djermoune, H.; Bonizzoni, L. Application of X-ray Fluorescence Spectroscopy to Provenance Studies of Algerian Archaeological Pottery. *X-ray Spectrom.* **2019**, *48*, 505–512. [CrossRef]
4. Bonizzoni, L. ED-XRF Analysis for Cultural Heritage: Is Quantitative Evaluation Always Essential? *J. Phys. Conf. Ser.* **2015**, *630*, 012001. [CrossRef]
5. Moioli, P.; Seccaroni, C. Analysis of Art Objects Using a Portable X-ray Fluorescence Spectrometer. *X-ray Spectrom.* **2000**, *29*, 48–52. [CrossRef]
6. Bonizzoni, L.; Galli, A.; Gondola, M.; Martini, M. Comparison between XRF, TXRF, and PXRF Analyses for Provenance Classification of Archaeological Bricks. *X-ray Spectrom.* **2013**, *42*, 262–267. [CrossRef]
7. Saleh, M.; Bonizzoni, L.; Orsilli, J.; Samela, S.; Gargano, M.; Gallo, S.; Galli, A. Application of Statistical Analyses for Lapis Lazuli Stone Provenance Determination by XRL and XRF. *Microchem. J.* **2020**, *154*, 104655. [CrossRef]
8. Bonizzoni, L.; Galli, A.; Milazzo, M. XRF Analysis without Sampling of Etruscan Depurata Pottery for Provenance Classification. *X-ray Spectrom.* **2010**, *39*, 346–352. [CrossRef]
9. Padilla, R.; Espen, P.V.; Torres, P.P.G. The Suitability of XRF Analysis for Compositional Classification of Archaeological Ceramic Fabric: A Comparison with a Previous NAA Study. *Anal. Chim. Acta* **2006**, *558*, 283–289. [CrossRef]

10. Schiavon, N.; de Palmas, A.; Bulla, C.; Piga, G.; Brunetti, A. An Energy-Dispersive X-ray Fluorescence Spectrometry and Monte Carlo Simulation Study of Iron-Age Nuragic Small Bronzes ("Navicelle") from Sardinia, Italy. *Spectrochim. Acta Part B At. Spectrosc.* **2016**, *123*, 42–46. [CrossRef]
11. Bottaini, C.; Mirão, J.; Figuereido, M.; Candeias, A.; Brunetti, A.; Schiavon, N. Energy Dispersive X-ray Fluorescence Spectroscopy/Monte Carlo Simulation Approach for the Non-Destructive Analysis of Corrosion Patina-Bearing Alloys in Archaeological Bronzes: The Case of the Bowl from the Fareleira 3 Site (Vidigueira, South Portugal). *Spectrochim. Acta Part B At. Spectrosc.* **2015**, *103–104*, 9–13. [CrossRef]
12. dos Santos, H.C.; Caliri, C.; Pappalardo, L.; Catalano, R.; Orlando, A.; Rizzo, F.; Romano, F.P. Real-Time MA-XRF Imaging Spectroscopy of the Virgin with the Child Painted by Antonello de Saliba in 1497. *Microchem. J.* **2018**, *140*, 96–104. [CrossRef]
13. Ruberto, C.; Mazzinghi, A.; Massi, M.; Castelli, L.; Czelusniak, C.; Palla, L.; Gelli, N.; Betuzzi, M.; Impallaria, A.; Brancaccio, R.; et al. Imaging Study of Raffaello's "La Muta" by a Portable XRF Spectrometer. *Microchem. J.* **2016**, *126*, 63–69. [CrossRef]
14. Vornicu, N.; Bibire, C.; Murariu, E.; Ivanov, D. Analysis of Mural Paintings Using in Situ Non-Invasive XRF, FTIR Spectroscopy and Optical Microscopy. *X-ray Spectrom.* **2013**, *42*, 380–387. [CrossRef]
15. True Versus Forged in the Cultural Heritage Materials: The Role of PXRF Analysis—Galli—2014—X-ray Spectrometry—Wiley Online Library. Available online: https://analyticalsciencejournals.onlinelibrary.wiley.com/doi/10.1002/xrs.2461 (accessed on 11 January 2022).
16. Bonizzoni, L.; Bruni, S.; Galli, A.; Gargano, M.; Guglielmi, V.; Ludwig, N.; Lodi, L.; Martini, M. Non-Invasive in Situ Analytical Techniques Working in Synergy: The Application on Graduals Held in the Certosa Di Pavia. *Microchem. J.* **2015**, *126*, 172–180. [CrossRef]
17. Idjouadiene, L.; Mostefaoui, T.A.; Naitbouda, A.; Djermoune, H.; Mechehed, D.E.; Gargano, M.; Bonizzoni, L. First Applications of Non-Invasive Techniques on Algerian Heritage Manuscripts: The LMUHUB ULAHBIB Ancient Manuscript Collection from Kabylia Region (Afniq n Ccix Lmuhub). *J. Cult. Herit.* **2021**, *49*, 289–297. [CrossRef]
18. Kokiasmenou, E.; Caliri, C.; Kantarelou, V.; Germanos Karydas, A.; Romano, F.P.; Brecoulaki, H. Macroscopic XRF Imaging in Unravelling Polychromy on Mycenaean Wall-Paintings from the Palace of Nestor at Pylos. *J. Archaeol. Sci. Rep.* **2020**, *29*, 102079. [CrossRef]
19. Tsatsouli, K.; Nikolaou, E. The Ancient Demetrias Figurines: New Insights on Pigments and Decoration Techniques Used on Hellenistic Clay Figurines. *Sci. Technol. Archaeol. Res.* **2017**, *3*, 341–357. [CrossRef]
20. Bonizzoni, L.; Galli, A.; Poldi, G.; Milazzo, M. In Situ Non-Invasive EDXRF Analysis to Reconstruct Stratigraphy and Thickness of Renaissance Pictorial Multilayers. *X-ray Spectrom.* **2007**, *36*, 55–61. [CrossRef]
21. Bertucci, M.; Bonizzoni, L.; Ludwig, N.; Milazzo, M. A New Model for X-ray Fluorescence Autoabsorption Analysis of Pigment Layers. *X-ray Spectrom.* **2010**, *39*, 135–141. [CrossRef]
22. Bonizzoni, L.; Colombo, C.; Ferrati, S.; Gargano, M.; Greco, M.; Ludwig, N.; Realini, M. A Critical Analysis of the Application of EDXRF Spectrometry on Complex Stratigraphies. *X-ray Spectrom.* **2011**, *40*, 247–253. [CrossRef]
23. Fiorini, C.; Gianoncelli, A.; Longoni, A.; Zaraga, F. Determination of the Thickness of Coatings by Means of a New XRF Spectrometer. *X-ray Spectrom.* **2002**, *31*, 92–99. [CrossRef]
24. Baumann, J.; Kayser, Y.; Kanngießer, B. Grazing Emission X-ray Fluorescence: Novel Concepts and Applications for Nano-Analytics. *Phys. Status Solidi* **2021**, *258*, 2000471. [CrossRef]
25. Baumann, J.; Grötzsch, D.; Scharf, O.; Kodalle, T.; Bergmann, R.; Bilchenko, F.; Mantouvalou, I.; Kanngießer, B. A Compact and Efficient Angle-Resolved X-ray Fluorescence Spectrometer for Elemental Depth Profiling. *Spectrochim. Acta Part B At. Spectrosc.* **2021**, *181*, 106216. [CrossRef]
26. Cesareo, R.; de Assis, J.T.; Roldán, C.; Bustamante, A.D.; Brunetti, A.; Schiavon, N. Multilayered Samples Reconstructed by Measuring Kα/Kβ or Lα/Lβ X-ray Intensity Ratios by EDXRF. *Nucl. Instrum. Methods Phys. Res. Sect. B Beam Interact. Mater. At.* **2013**, *312*, 15–22. [CrossRef]
27. Karimi, M.; Amiri, N.; Tabbakh Shabani, A.A. Thickness Measurement of Coated Ni on Brass Plate Using Kα/Kβ Ratio by XRF Spectrometry. *X-ray Spectrom.* **2009**, *38*, 234–238. [CrossRef]
28. Gargano, M.; Galli, A.; Bonizzoni, L.; Alberti, R.; Aresi, N.; Caccia, M.; Castiglioni, I.; Interlenghi, M.; Salvatore, C.; Ludwig, N.; et al. The Giotto's Workshop in the XXI Century: Looking inside the "God the Father with Angels" Gable. *J. Cult. Herit.* **2019**, *36*, 255–263. [CrossRef]
29. Orsilli, J.; Galli, A.; Bonizzoni, L.; Caccia, M. More than XRF Mapping: STEAM (Statistically Tailored Elemental Angle Mapper) a Pioneering Analysis Protocol for Pigment Studies. *Appl. Sci.* **2021**, *11*, 1446. [CrossRef]
30. Galli, A.; Caccia, M.; Caglio, S.; Bonizzoni, L.; Castiglioni, I.; Gironda, M.; Alberti, R.; Martini, M. An Innovative Protocol for the Study of Painting Materials Involving the Combined Use of MA-XRF Maps and Hyperspectral Images. *Eur. Phys. J. Plus* **2021**, *137*, 22. [CrossRef]
31. Bottaini, C.; Mirão, J.; Candeias, A.; Catarino, H.; Silva, R.J.; Brunetti, A. Elemental Characterisation of a Collection of Metallic Oil Lamps from South-Western al-Andalus Using EDXRF and Monte Carlo Simulation. *Eur. Phys. J. Plus* **2019**, *134*, 365. [CrossRef]
32. Trojek, T. Iterative Monte Carlo Procedure for Quantitative X-ray Fluorescence Analysis of Copper Alloys with a Covering Layer. *Radiat. Phys. Chem.* **2020**, *167*, 108294. [CrossRef]

33. Fermo, P.; Andreoli, M.; Bonizzoni, L.; Fantauzzi, M.; Giubertoni, G.; Ludwig, N.; Rossi, A. Characterisation of Roman and Byzantine Glasses from the Surroundings of Thugga (Tunisia): Raw Materials and Colours. *Microchem. J.* **2016**, *129*, 5–15. [CrossRef]

34. Bonizzoni, L.; Bruni, S.; Guglielmi, V.; Milazzo, M.; Neri, O. Field and Laboratory Multi-Technique Analysis of Pigments and Organic Painting Media from an Egyptian Coffin (26th Dynasty). *Archaeometry* **2011**, *53*, 1212–1230. [CrossRef]

35. Nuevo, M.J.; Martín Sánchez, A. Application of XRF Spectrometry to the Study of Pigments in Glazed Ceramic Pots. *Appl. Radiat. Isot.* **2011**, *69*, 574–579. [CrossRef] [PubMed]

36. Karydas, A.G. Application of a Portable XRF Spectrometer for the Non-Invasive Analysis of Museum Metal Artefacts. *Ann. Chim.* **2007**, *97*, 419–432. [CrossRef]

37. Ruvalcaba Sil, J.L.; Ramírez Miranda, D.; Aguilar Melo, V.; Picazo, F. SANDRA: A Portable XRF System for the Study of Mexican Cultural Heritage. *X-ray Spectrom.* **2010**, *39*, 338–345. [CrossRef]

38. A Novel Portable XRF Spectrometer with Range of Detection Extended to Low-Z Elements—Migliori—2011—X-ray Spectrometry—Wiley Online Library. Available online: https://analyticalsciencejournals.onlinelibrary.wiley.com/doi/10.1002/xrs.1316 (accessed on 11 January 2022).

39. Campos, P.H.O.V.; Appoloni, C.R.; Rizzutto, M.A.; Leite, A.R.; Assis, R.F.; Santos, H.C.; Silva, T.F.; Rodrigues, C.L.; Tabacniks, M.H.; Added, N. A Low-Cost Portable System for Elemental Mapping by XRF Aiming in Situ Analyses. *Appl. Radiat. Isot.* **2019**, *152*, 78–85. [CrossRef] [PubMed]

40. Vittiglio, G.; Bichlmeier, S.; Klinger, P.; Heckel, J.; Fuzhong, W.; Vincze, L.; Janssens, K.; Engström, P.; Rindby, A.; Dietrich, K.; et al. A Compact μ-XRF Spectrometer for (In Situ) Analyses of Cultural Heritage and Forensic Materials. *Nucl. Instrum. Methods Phys. Res. Sect. B Beam Interact. Mater. At.* **2004**, *213*, 693–698. [CrossRef]

41. Galli, A.; Bonizzoni, L.; Martini, M.; Sibilia, E. Archaeometric Study of Fictile Tubes from Three Churches in Milan. *Appl. Phys. A* **2008**, *92*, 117–121. [CrossRef]

42. Ametek, Choosing the Anode Material. 2021. Available online: https://www.amptek.com/-/media/ametekamptek/documents/products/choosing-the-anode-material-in-an-X-ray-tube.pdf (accessed on 30 November 2021).

43. Ametek, Mini-X User Manual. 2021. Available online: https://www.amptek.com/-/media/ametekamptek/documents/products/user-manuals/mini-x2-user-manual-rev-b2.pdf (accessed on 11 January 2022).

44. Ametek, Amptek Mini-X X-ray Tube Application Note Filters on an X-ray Tube. 2021. Available online: https://www.amptek.com/-/media/ametekamptek/documents/products/filters-application-note.pdf (accessed on 30 November 2021).

45. Bonizzoni, L.; Galli, A.; Milazzo, M. Direct Evaluation of Self-Absorption Effects in Dark Matrices by Compton Scattering Measurements. *X-ray Spectrom.* **2000**, *29*, 443–448. [CrossRef]

46. Solidworks, Solidworks Desktop 3D CAD. Available online: http://dl-ak.solidworks.com/nonsecure/training/TOC_PMT2105-ENG_MLD2021.pdf (accessed on 11 January 2022).

47. Neelmeijer, C.; Brissaud, I.; Calligaro, T.; Demortier, G.; Hautojärvi, A.; Mäder, M.; Martinot, L.; Schreiner, M.; Tuurnala, T.; Weber, G. Paintings—A Challenge for XRF and PIXE Analysis. *X-ray Spectrom.* **2000**, *29*, 101–110. [CrossRef]

48. Pigment Compendium. Available online: https://www.routledge.com/pigment-compendium/eastaugh-walsh-chaplin-siddall/p/book/9780750689809 (accessed on 11 January 2022).

49. Raw Sienna, ColourLex. Available online: https://colourlex.com/project/raw-sienna/ (accessed on 1 December 2021).

50. Beckhoff, B.; Kanngieser, B.; Langhoff, N.; Wedell, R.; Wolff, H. *Handbook of Practical X-ray Fluorescence Analysis*, 2006 ed.; Springer Nature: Basingstoke, UK, 2006; ISBN 978-3-662-49601-5.

51. Bonizzoni, L.; Caglio, S.; Galli, A.; Poldi, G. Comparison of Three Portable EDXRF Spectrometers for Pigment Characterization. *X-Ray Spectrom.* **2010**, *39*, 233–242. [CrossRef]

52. Marengo, E.; Aceto, M.; Robotti, E.; Liparota, M.C.; Bobba, M.; Pantò, G. Archaeometric Characterisation of Ancient Pottery Belonging to the Archaeological Site of Novalesa Abbey (Piedmont, Italy) by ICP–MS and Spectroscopic Techniques Coupled to Multivariate Statistical Tools. *Anal. Chim. Acta* **2005**, *1–2*, 359–375. [CrossRef]

53. D'Alessandro, A.; Lucarelli, F.; Mandò, P.; Marcazzan, G.; Nava, S.; Prati, P.; Valli, G.; Vecchi, R.; Zucchiatti, A. Hourly Elemental Composition and Sources Identification of Fine and Coarse PM10 Particulate Matter in Four Italian Towns. *J. Aerosol Sci.* **2003**, *34*, 243–259. [CrossRef]

54. R Core Team. R: A Language and Environment for Statistical Computing, R Foundation for Statistical Compu-Ting. 2021. Available online: https://www.r-project.org/ (accessed on 13 December 2021).

55. Lê, S.; Josse, J.; Husson, F. FactoMineR: An R Package for Multivariate Analysis. *J. Stat. Softw.* **2008**, *25*, 1–18. [CrossRef]

56. Weiner, J. Pca3d: Three Dimensional PCA Plots. R Package Version 0.10.2. 2020. Available online: https://cran.r-project.org/package=pca3d (accessed on 11 January 2022).

57. Aruga, R. Closure of Analytical Chemical Data and Multivariate Classification. *Talanta* **1998**, *47*, 1053–1061. [CrossRef]

58. Kregsamer, P. *QXAS—Quantitative X-ray Analysis System (User's Manual and Guide to X-ray Fluorescence Technique)*; IAEA Computer Manual Series No. 21 (IAEA/CMS/21/CD); IAEA: Vienna, Austria, 2009.

Article

New Insights on the Stradivari "Coristo" Mandolin: A Combined Non-Invasive Spectroscopic Approach

Francesca Volpi [1,2], Giacomo Fiocco [1], Tommaso Rovetta [1], Claudia Invernizzi [1], Michela Albano [1,2], Maurizio Licchelli [3] and Marco Malagodi [1,2,*]

[1] Arvedi Laboratory of Non-Invasive Diagnostics, CISRiC, University of Pavia, Via bell'Aspa 3, 26100 Cremona, Italy; francesca.volpi@unipv.it (F.V.); giacomo.fiocco@unipv.it (G.F.); tommaso.rovetta@unipv.it (T.R.); claudia.invernizzi@unipv.it (C.I.); michela.albano@unipv.it (M.A.)

[2] Department of Musicology and Cultural Heritage, University of Pavia, Corso Garibaldi, 178, 26100 Cremona, Italy

[3] Department of Chemistry, University of Pavia, Via Taramelli 12, 27100 Pavia, Italy; maurizio.licchelli@unipv.it

* Correspondence: marco.malagodi@unipv.it

Abstract: In this work, one of the two existing mandolins made by Antonio Stradivari has been investigated for the first time, as a rare exemplar of the lesser-known class of plucked string instruments. The mandolin was studied by non-invasive reflection Fourier transformed infrared (FT-IR) spectroscopy and X-ray fluorescence (XRF) on different areas previously selected by UV-induced fluorescence imaging. The analytical campaign was aimed at (i) identifying the materials used by Stradivari in the finishing of the mandolin, (ii) comparing these materials with those traditionally used in violin making, and (iii) increasing the knowledge of materials and techniques applied by Stradivari in the rare production of plucked string instruments. The combined spectroscopic approach allowed us to hypothesize original materials and finishing procedures similar to those used in violin making: a possible sizing treatment of the wood with protein-based materials and silicates, externally coated with an oil–resin varnish. XRF results were essential to support FT-IR findings and to detect possible iron-based pigments in the finishing layers. Moreover, it permitted us to distinguish original areas from the restored areas, including the purflings on the top plate and the varnished area on the treble side of the mandolin for which the originality was assumed.

Keywords: Stradivari; musical instrument; mandolin; varnish; coatings; multi-layered structure; XRF; reflection FT-IR; spectroscopy

Citation: Volpi, F.; Fiocco, G.; Rovetta, T.; Invernizzi, C.; Albano, M.; Licchelli, M.; Malagodi, M. New Insights on the Stradivari "Coristo" Mandolin: A Combined Non-Invasive Spectroscopic Approach. *Appl. Sci.* **2021**, *11*, 11626. https://doi.org/10.3390/app112411626

Academic Editor: Wolfgang Elsaesser

Received: 30 October 2021
Accepted: 2 December 2021
Published: 7 December 2021

Publisher's Note: MDPI stays neutral with regard to jurisdictional claims in published maps and institutional affiliations.

1. Introduction

In recent decades, research efforts have dramatically advanced the knowledge of historical musical instruments in terms of finishing techniques and analytical approaches applied to characterize the utilized materials [1,2], focusing attention on bowed string instruments of the Cremonese tradition between the 17th and 18th centuries. The most famous luthier of this period is the renowned Master Antonio Stradivari (Cremona, 1644–1737), who left numerous excellent musical instruments and several traces of his works, including drawings, molds, and tools [3], but not written sources of his finishing techniques. For this reason, the role of scientific investigation is crucial to characterize the finish of the instruments, which was revealed to be a complex multi-layered coating system where each layer is likely representative of a specific step of the finishing procedure.

Regarding the composition of the excellent Stradivari's varnish, chemical analysis performed on several violins made by the great Master, dated from 1679 to 1724, often identified a mixture of siccative oil and natural resins [4–7]. In some cases, the varnish was enriched with low concentrations of mineral pigments, mostly Fe-based earth or organic colorants, to give a slightly colorful appearance while keeping the transparency and glossy effect [8–11]. Bone ash and pumice have been also suggested as minerals added to stabilize and increase the rate of drying of the varnish [12]. Under the varnish, a ground layer was often detected and described as a mixture of protein-based materials with a dispersion of inorganic fillers, such as silicates or sulfates [13]. Moreover, the presence of a pre-treatment of the wood has been suggested by several authors, although it was barely detected and identified [14–16].

The contribution of Antonio Stradivari to instrument making is not only limited to bowed string instruments, i.e., violin, viola, and cello, but also includes plucked string instruments such as guitars, lutes, harps, and mandolins. His wide production is confirmed by tens of drawings and molds of different kinds of musical instruments currently conserved in the Museo del Violino in Cremona, Italy [3]. Among the plucked string instruments of Stradivari, only two mandolins are known to have survived: the oldest, named "The Cutler-Callen", made in 1680 and conserved in the National Music Museum in Vermillion, South Dakota, USA [16]; and the "Coristo" mandolin that was made between 1700 and 1710, currently conserved in the Museo del Violino in Cremona (Figure 1; high-resolution images are reported in Figures S1–S6 in Supplementary Information). The latter extraordinary example of the Stradivari musical instrument is the object of this study.

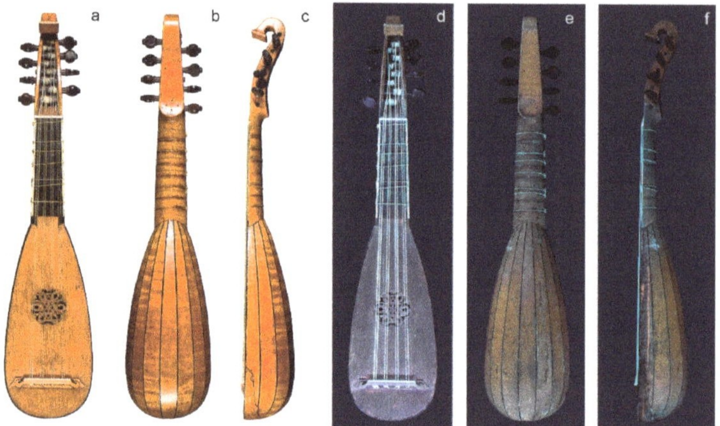

Figure 1. Images in visible light (**a–c**), and UV-induced fluorescence (**d–f**) of the front, the back, and the right treble side of the "Coristo" mandolin.

From historical sources, we know that the mandolin probably derived from an ancient Arab instrument similar to a small lute, and it succeeded in Italy principally over the 17th and 18th centuries due to an important production of music made on purpose for this type of instrument [17,18]. Concerning the materials used to finish a mandolin, though, little information is available [19]. In this view, for the first time, the analytical results on the "Coristo" mandolin made by Stradivari are here presented, using a combined approach of three non-invasive techniques: UV-induced fluorescence imaging, reflection Fourier transformed infrared (FT-IR) spectroscopy, and X-ray fluorescence (XRF). These complementary techniques were selected based on the capability of UV, IR, and X-ray radiations to obtain information from different organic and inorganic materials, allowing us to non-invasively characterize the coating system. The goals of the study were to: (i) identify the materials used by Stradivari in the finishing processes of the mandolin, (ii) compare these materials with those used in violin finishing, and (iii) increase the knowledge of materials and techniques applied by the Master in the less-explored class of plucked string instruments.

2. Materials and Methods

2.1. The "Coristo" Mandolin

The "Coristo" mandolin is currently owned by a private collector. It belonged to an American museum, which sent it for sale by auction in the 1970s. As described by the collector Beare, when the mandolin was sold, "it showed a neglected state of conservation, it was dirty, showing several—not original—added layers of varnish. It was successively restored, and the added varnishes were removed to show again the typical Stradivari varnish" [18]. Based on a preliminary observation of the surface with visible and UV light and the investigation of the structural features of the mandolin through X-ray radiography (not reported in this paper), a good state of conservation characterized the instrument when it was analyzed.

2.2. The Methodological Approach

The deterioration of an ancient musical instrument can cause varnish detachments or thinning of the coating, especially in correspondence with worn-out areas produced by contact with the musician, exposing the inner layers. To approach the non-invasive stratigraphic investigation of the Stradivari's mandolin, without any sampling, the analyses were performed on both well-conserved and worn-out areas to give insight into the materials and methods used by the Master. A preliminary observation of the instrument surface was performed in visible light while UV-induced fluorescence imaging was used to highlight variations in the optical properties, likely due to chemical inhomogeneities of the materials and potentially the presence of additional layers [20,21], or even the thinning of the existing one. Finally, the selected areas of interest were investigated by two spectroscopic techniques, namely X-ray fluorescence (XRF) [22,23] and reflection Fourier transform infrared (FT-IR) [5,24,25], to achieve both elemental and molecular information, respectively.

A Nikon D4 full-frame digital camera (Minato, Tokyo, Japan) equipped with a 50 mm f/1.4 Nikkor objective was used. Visible light images (f/11 and ISO 100) were obtained using a Softbox LED lamp, while UV-induced fluorescence ones (30 s exposure time, f/11, and ISO 400) were collected by illuminating the samples with two Philips TL-D 36W BBL IPP low-pressure Hg tubes (emission peak at 365 nm). In order to highlight and label areas of interest on the surface of the instrument, the UV Analyzer software, self-developed by the Arvedi Laboratory at the University of Pavia, was used [26–28].

Reflection FT-IR spectra were recorded using the Alpha portable spectrometer (Bruker Optics, Etringen, Germany; Billerica, MA, USA) equipped with a SiC globar source, a permanently aligned RockSolid interferometer with gold mirrors, and a DLaTGS detector. Measurements were performed at a working distance of 15 mm by an external reflectance module with an optical layout of $23°/23°$. Pseudo-absorbance spectra ($\log(1/R)$; R = reflectance) were acquired between 7500 and 375 cm^{-1} with a spectral resolution of 4 cm^{-1} and acquisition time of 1 min. A gold flat mirror was used as the background. Reflection spectra were transformed to absorbance spectra by applying the Kramers–Kronig (KK) algorithm, included in the OPUS 7.2 software package, and the mid-IR spectral range is exhibited in the figures. The FT-IR analytical spots acquired on the top plate, shell, and treble side of the mandolin are shown in Figure S7 (green circles) in Supplementary Information.

X-ray fluorescence spectra were acquired by a portable energy-dispersive XRF spectrometer ELIO (Bruker Corporation, Billerica, MA, USA). The excitation source works with an Rh anode and the beam is collimated to a spot diameter on the sample surface of about 1.3 mm. XRF measurements were carried out by fixing the tube voltage at 40 kV and the tube current at 80 µA for a measured time of 480 s (8 min) and setting acquisition channels at 2048. The data were processed by ELIO 1.6.0.29 software (Bruker Corporation, Billerica, MA, USA). To qualitatively estimate the abundance of the elements, the net area count of each element was normalized by dividing the Kα peaks (except for Ba and Pb where Lα was used) by the averaged net area counts of the coherent scattering peak of Rh. It is worth clarifying that due to the working conditions used in this study, i.e., air as medium or the presence of multi-layered substrates with variable thicknesses, the detection of light elements such as silicon, sulfur, and phosphor may result underestimated because their secondary radiations can be severely attenuated by the air and/or any dense material superimposed [10,29]. At any event, even though a fully semiquantitative analysis was not possible, the use of the same geometry, voltage, and current conditions for different points of analysis allowed a reasonable qualitative comparison. The XRF analytical spots acquired on the top plate, shell, and treble side of the mandolin are shown in Figure S7 (red squares) in Supplementary Information.

3. Results and Discussion

3.1. Selection of the Areas of Analysis

Based on the UV-induced fluorescence colors of the mandolin surface in Figure 1, seven areas of interest were highlighted through the UV Analyzer software shown in Figure 2. In the top plate, two UV-induced fluorescence colors were observed: a fairly homogeneous blue hue over the majority of the plate and some light-brown spots. These two areas were identified by the UV Analyzer in Figure 2a as A and B, respectively. The well-conserved varnished area on the shell of the mandolin showed a yellow-brown UV-induced fluorescence color, while worn areas displayed a blue-greenish fluorescence, respectively identified as C and D areas in Figure 2b. In correspondence with the treble side of the mandolin (Figure 1f), an orange UV fluorescent area appeared, which was identified as area E in Figure 2c. The purflings on the top plate and shell which showed a dark UV-induced fluorescence color were identified as F in Figure 2.

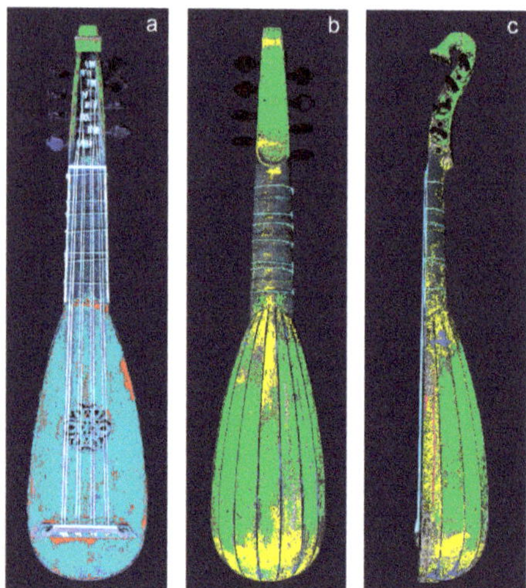

Figure 2. Images of the top plate (**a**), shell (**b**), and treble side (**c**) processed by the UV Analyzer. The colors correspond to different areas of interest according to the UV-induced fluorescence color: area A in light-blue, area B in red, area C in green, area D in yellow, area E in magenta, and area F (purflings) in black.

3.2. Reflection FT-IR Spectroscopic Analysis

All the reflection FT-IR spectra acquired on the top plate of the mandolin (area A in Figure 2) show similar features, where the prevalent signals derive from the wooden substrate. Typical spectral features of wood include a broad OH bond vibration band at around 3350 cm^{-1}, weak and poorly resolved CH stretching vibrations around 2900 and 2800 cm^{-1}, and the bands at 1720 cm^{-1} attributed to unconjugated C=O. In addition, wood shows intense multiple signals between 1150 and 1050 cm^{-1} assigned to the stretching of C-O bonds and glucose ring stretches in cellulose [30,31]. The bands attributed to the wood are marked with black rhombus in the KK-transformed spectrum in Figure 3a. In addition, poorly resolved features at 1660–1650 cm^{-1} and 1550 cm^{-1} appeared in the KKT spectra (marked with a triangle in Figure 3a,b), suggesting the presence of proteinaceous compounds at 1660 cm^{-1} (amide I) and 1550 cm^{-1} (amide II) [13,32]. In the pseudo-absorbance spectra, the Restrahlen band at 1015 cm^{-1} (marked with an asterisk in Figure 3a,b) suggests the presence of silicates [33,34]. Furthermore, the detection of sharp CH stretching bands at 2920 cm^{-1} and 2850 cm^{-1}, particularly intense in the right and bottom areas of the top plate, along with CH bending bands between 1400 and 1350 cm^{-1}, and a weak and not-resolved C=O signal around 1710 cm^{-1} appearing as a shoulder in the KKT spectra of Figure 3, could possibly suggest the additional presence of organic materials. However, a clear identification of this class of compounds was not permitted due to poor intensity of the bands and the predominance of wood signals in the spectral profile. Some KKT spectra collected on area B, highlighted in red in Figure 2, display sharp peaks at 1260 cm^{-1} and around 800 cm^{-1} (marked with circles in Figure 3b), which may correspond to Si-CH$_3$ stretching bonds [35], presumably related to a modern silicone-based product used to superficially treat and polish the wood. Organosilicon compounds were also revealed on the "San Lorenzo" violin made in 1718 by Stradivari as a residue of a contemporary polishing treatment [11].

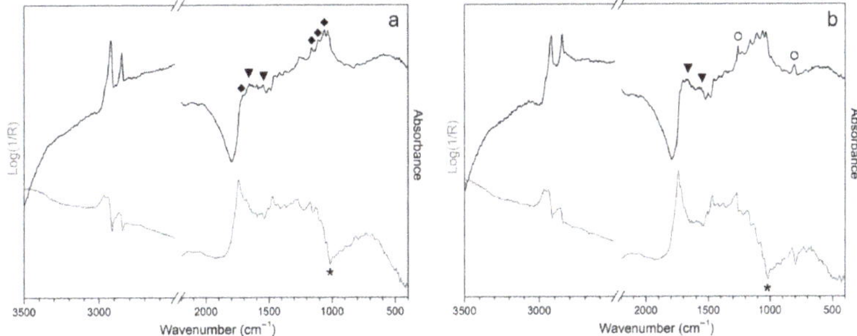

Figure 3. Reflection FT-IR spectra in pseudo-absorbance (grey line) and after KK transform (black line) acquired on area A (**a**) and B (**b**) of the top plate as identified by the UV Analyzer in Figure 2a. Markers for proteins (triangle), wood (black rhombus), silicates (asterisk), and silicon-based material (circle) are displayed.

The FT-IR spectrum representative of the well-conserved areas of the shell (area C in Figure 2b) exhibits the signals of an oil–resin varnish, likely linseed oil in mixture with a terpenic resin [6,28,31], with typical CH stretching bands at 2920 cm^{-1} and 2850 cm^{-1}, CH bending at 1462, 1376, and 1385 cm^{-1}, and a broad C=O stretch around 1715 cm^{-1} in the KKT spectrum (Figure 4a) [31,36,37]. The carbonyl band is broadened due to a double contribution from the ester groups (oil) around 1740 cm^{-1} and the acid groups (resin) around 1715–1690 cm^{-1}, which broadens with degradation and oxidation [38–40]. Interestingly, the spectrum acquired on the worn-out areas of the shell (area D in Figure 2b) shows signals increasing around 1650 and 1550 cm^{-1}, in the KK-transformed spectrum, and around 1015 cm^{-1} in the pseudo-absorbance spectrum, marked in Figure 4b, previously observed on the top plate, respectively ascribable to protein-based products and silicates used as sizing treatment. The detection and identification of proteins and silicates on wood is still a discussed topic in scientific literature. However, in accordance with many studies, the detection of these compounds in an area where the varnish was almost totally detached or significantly thinned by the use of the mandolin during its history seems to support fairly well the hypothesis of a wood sizing treatment applied before varnishing. Therefore, it is reasonable to assume that an oil–resin varnish was applied on the surface of the mandolin and, where the varnish layer is consumed, the underneath treated wooden surface is revealed.

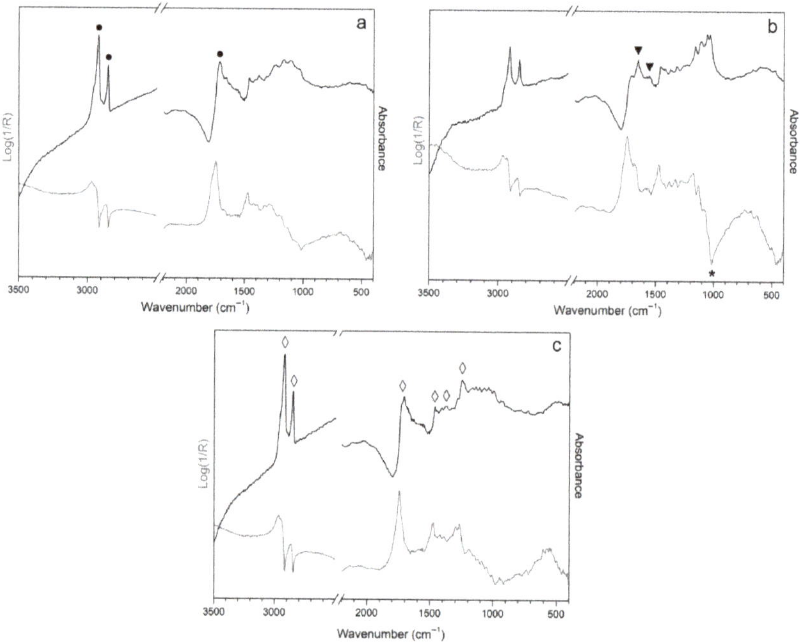

Figure 4. Reflection FT-IR spectra in pseudo-absorbance (grey line) and after KK transform (black line) acquired on C (**a**), D (**b**), and E (**c**) areas of the shell as identified by the UV Analyzer in Figure 2. Markers for oil–resin varnish (circle), proteins (triangle), silicates (asterisk), and shellac (white rhombus) are displayed.

Finally, area E, highlighted in magenta in Figure 2c, corresponds fairly well with the reference spectrum of shellac (Figure 4c). This animal resin, excreted from lac beetles, shows typical CH stretching bands at 2920 cm^{-1} and 2857 cm^{-1}, CH$_2$ scissoring bands at 1465 cm^{-1}, a C=O stretching doublet at 1735 cm^{-1} and 1715 cm^{-1} due to the ester and acid groups, respectively [32], and the characteristic intense peak at 1252 cm^{-1} due to C-O stretches of ester bonds [4]. Shellac was used in Italy as early as the 18th century, and it is also commonly used in mixture with alcohol in the so-called French polish technique in musical instruments conservation to enhance the aesthetic features of the varnish [41].

3.3. XRF Results

A qualitative evaluation of XRF data acquired on different areas of the mandolin is reported in Table 1. The results obtained on the top plate and the shell reveal high counts of potassium (K) and calcium (Ca). These elements are naturally found in wood, as displayed in Figure 5a; however, their increment along with not negligible counts of sulphur (S) and phosphorus (P) and the signals of proteinaceous material obtained from FT-IR analysis (Section 3.2) could imply the presence of protein-based wood treatments, such as animal glue and/or potassium/calcium caseinate, which could have been applied before varnishing the surface [10]. The addition of bone ash or pumice to the varnish, as proposed by other authors [12], could explain the increment of Ca, K, and P, but FT-IR analysis did not reveal hydroxyapatite, which is the main constituent of bone, while the presence of silicate, also contained in pumice, seemed to be likely located within the inner layer rather than in the varnish.

Table 1. List of the elements detected by XRF in different areas of the mandolin, listed according to their normalized peak area counts, from the most (>10) to the least (<1) abundant. The sign "-" in the table means "no elements".

Area of the Mandolin	>10	1–10	<1
Top plate	-	K, Ca	Si, P, S, Ti, Cr, Mn, Fe, Ni, Cu, Zn, Sr, Pb
Shell	-	K, Ca	Si, P, S, Ti, Fe, Cu, Pb
Restoration material on the treble side	-	Ca, Fe, Zn	S, K, Cr, Mn, Ni, Cu, Sr, Ba, Pb
Purfling on the shell (SP)	Fe	K, Ca	Si, P, S, Ti, Cr, Mn, Cu, Zn, Sr, Pb
Fingerboard	Fe	K, Ca	Si, P, S, Ti, Mn, Cu, Zn, Sr, Pb
Purfling on the top plate (TP)	-	K, Ca, Fe	Si, P, S, Ti, Cr, Mn, Ni, Cu, Zn, Br, Sr, Pb

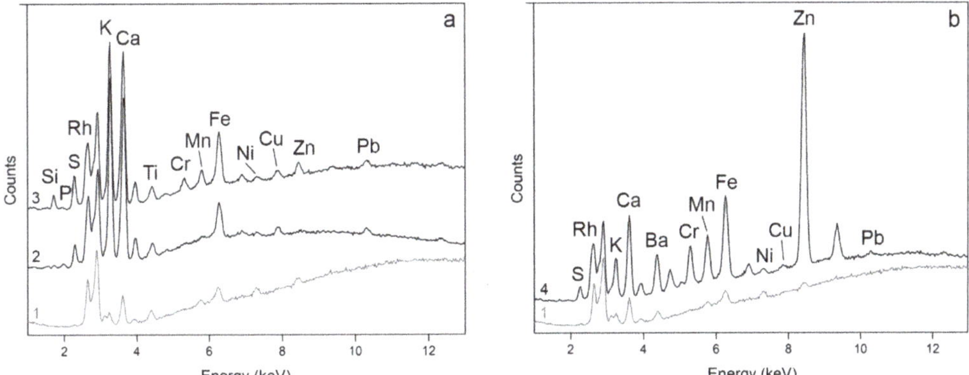

Figure 5. XRF representative spectra acquired on the top plate and the shell (**a**) and on the treble side of the mandolin (**b**). Numbers refer to different XRF spectra collected on Picea abies reference wood (1), areas A (2), C (3), and E (4) as identified through the UV Analyzer in Figure 2.

The detection of manganese (Mn) in the top plate, in combination with iron (Fe), may suggest the use of umber earth pigment, which is composed mainly of Fe and Mn oxides [41]. Conversely, no Mn was detected on the well-conserved areas of the shell. The presence of zinc (Zn), nickel (Ni), and chromium (Cr) was detected only on the top plate and not on the shell and could be related to pollutants. Silicon (Si) was revealed in each point of analysis, both on the top plate and the shell. Although the related counts may be severely underestimated, the occurrence of silicon can be ascribable to silicate-based particles, as highlighted also by the results obtained through reflection FT-IR and described in Section 3.2 (a schematic overview of the results is reported in Table S1 in Supplementary Information). Lead (Pb) was always detected and its presence could imply, according to the literature, the use of oils or oil–resin products, since Pb-based siccative agents, e.g., massicot (PbO) and white lead (2PbCO$_3$·Pb(OH)$_2$), were added to accelerate the drying of oil [42–44].

The XRF data acquired on area E of the treble side of the mandolin are displayed in Figure 5b, showing remarkable differences from those acquired either on the top plate or the shell of the mandolin. Zn together with barium (Ba) could suggest the presence of lithopone, a mixture of ZnS and BaSO$_4$ [45] used since the early 19th century in cold-water paints, although this hypothesis was not supported by FT-IR. Furthermore, in the same spot, Fe, Cr, and Mn increase in counts, likely suggesting the addition of a pigment in the finishing treatment. These results enabled us to hypothesize the application of an additional colored and pigmented coat on worn areas to resemble the original appearance [11]. To deepen the investigation of the complex stratigraphies present in this type of work of art, i.e., a historical musical instrument, high-performance techniques such as XRF mapping, micro-XRD, and micro-tomography at synchrotron facilities would be considered along with grazing-angle methods.

The purfling analyzed on the shell (SP), as well as the fingerboard (FB), show Fe counts one order of magnitude greater than all the other analyzed areas. In addition, S, Cu, and Zn signals increase, suggesting that wood was probably dyed with iron–gall ink, which contains mineral components such as iron sulfate and minor portions of copper sulfate and traces of zinc impurities. In ancient times, indeed, dyeing wood was a very widespread practice to obtain substitutes for the rare, hard-to-work, and expensive ebony [19,44,46,47]. When compared to SP and FB in Figure 6a, the purflings on the top plate (TP_1 and TP_2, respectively) show lower counts of Fe, Cu, and Zn, reported as Fe/Cu and Cu/Zn ratios in Figure 6b, likely due to the use of naturally black wood such as ebony. Moreover, bromide (Br) was detected in TP_1 and TP_2, and this occurrence led us to hypothesize the application of methyl bromide during a fumigation pest treatment, as typical from the 1930s to modern times [48]. In the "Coristo" mandolin, the detection of Br occurred only in the purflings on the top plate; hence it is reasonable to assume that only those decorative elements were replaced with modern fumigated wood.

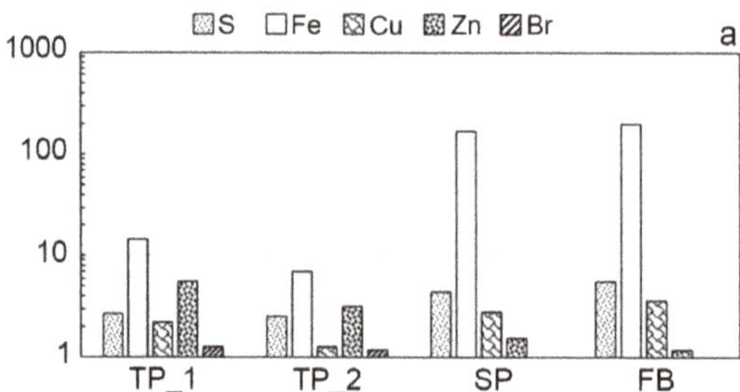

Figure 6. Estimation of elements detected by XRF on the top plate purflings (TP_1 and TP_2), on the shell purflings between the staves (SP), and on the fingerboard (FB). The values correspond to the net area counts of five different peaks (S-, Fe-, Cu-, Zn-, and Br-Kα) divided by net area counts of the peak of Rh-Kα, and they are displayed in a logarithmic scale (**a**). Elemental ratios calculated from net peak area counts (**b**).

4. Conclusions

This study presents the first analytical results obtained on the "Coristo" mandolin, one of the two existing mandolins made by Antonio Stradivari, through the combination of XRF and reflection FT-IR spectroscopic non-invasive techniques. The results revealed that the shell of the instrument was finished with an oil–resin varnish, as already identified in previous research on violins and other bowed string instruments made by Stradivari, while the top plate was left probably unvarnished, although FT-IR spectra revealed organic materials difficult to characterize. A proteinaceous sizing with silicates as inorganic fraction was detected both on the top plate and the worn areas of the shell, suggesting a treatment of the wood, in accordance with the findings in other Stradivari instruments. In addition, residues of posthumous maintenance treatments were detected in some spots on the top plate and on the treble side, where shellac varnish was used along with fillers and iron-based pigments to resemble the original surface. Regarding the black purflings on the top plate and between the staves of the shell, we characterized two different elemental compositions. The purflings on the shell were made of wood dyed with iron–gall ink rather than the more expensive and exotic ebony. The purflings on the top plate were probably substituted in modern times with naturally black wood, likely ebony, fumigated with methyl bromide on the basis of the detection of bromide limited to this part of the mandolin. According to the non-invasive results achieved so far, we can conclude that the materials and finishing procedures used in the Stradivari's mandolin were the same as those used in violins; therefore, the typical features of the workshop are also recognizable in plucked string instruments.

Supplementary Materials: The following are available online at https://www.mdpi.com/2076-341 7/11/24/11626/s1, Figure S1: Image at high resolution in visible light of the front of the "Coristo" mandolin, Figure S2: Image at high resolution in visible light of the shell of the "Coristo" mandolin, Figure S3: Image at high resolution in visible light of the treble side of the "Coristo" mandolin, Figure S4: Image at high resolution in UV light of the front of the "Coristo" mandolin, Figure S5: Image at high resolution in UV light of the shell of the "Coristo" mandolin, Figure S6: Image at high resolution in UV light of the treble side of the "Coristo" mandolin, Figure S7: UV images of the front (a), the shell (b), ande the treble side (c) with the FT-IR and XRF analytical spots market with green circles and red squares, respectively, Table S1: Summary of the findings achieved with UV-vis Imaging, Reflection FT-IR and XRF.

Author Contributions: Conceptualization, F.V., G.F., T.R. and M.M.; methodology, F.V., G.F., T.R. and C.I.; formal analysis, G.F., T.R., C.I. and M.A.; data curation, F.V., G.F., T.R. and C.I.; writing—original draft preparation, F.V.; writing—review and editing, F.V., G.F., T.R., C.I., M.A., M.M. and M.L.; validation M.M. and M.L. All authors have read and agreed to the published version of the manuscript.

Funding: This research received no external funding.

Institutional Review Board Statement: Not applicable.

Informed Consent Statement: Not applicable.

Data Availability Statement: Supporting data can be provided under request.

Acknowledgments: A special acknowledgment goes to Fondazione Antonio Stradivari Museo del Violino and to the Conservator Fausto Cacciatori.

Conflicts of Interest: The authors declare no conflict of interest.

References

1. Fiocco, G.; Invernizzi, C.; Rovetta, T.; Albano, M.; Malagodi, M.; Davit, P.; Gulmini, M. Surfing through the coating system of historic bowed instruments: A spectroscopic perspective. *Spectrosc. Eur.* **2021**, *33*, 19–22. [CrossRef]
2. Invernizzi, C.; Fiocco, G.; Iwanicka, M.; Targowski, P.; Piccirillo, A.; Vagnini, M.; Licchelli, M.; Malagodi, M.; Bersani, D. Surface and interface treatments on wooden artefacts: Potentialities and limits of a non-invasive multi-technique study. *Coatings* **2021**, *11*, 29. [CrossRef]
3. Cacciatori, F. *Antonio Stradivari, Disegni, Modelli e Forme–Catalogo Dei Reperti Delle Collezioni Civiche Liutarie del Comune di Cremona, Cremona*; Fondazione Museo del Violino Antonio Stradivari: Cremona, Italy, 2016; ISBN 88-9091-795-0.
4. Invernizzi, C.; Daveri, A.; Rovetta, T.; Vagnini, M.; Licchelli, M.; Cacciatori, F.; Malagodi, M. A multi-analytical non-invasive approach to violin materials: The case of Antonio Stradivari 'Hellier' (1679). *Microchem. J.* **2016**, *124*, 743–750. [CrossRef]
5. Invernizzi, C.; Fiocco, G.; Iwanicka, M.; Kowalska, M.; Targowski, P.; Blümich, B.; Rehorn, C.; Gabrielli, V.; Bersani, D.; Licchelli, M.; et al. Non-invasive mobile technology to study the stratigraphy of ancient Cremonese violins: OCT, NMR-MOUSE, XRF and reflection FT-IR spectroscopy. *Microchem. J.* **2020**, *155*, 104754. [CrossRef]
6. Echard, J.P.; Benoit, C.; Peris-Vicente, J.; Malecki, V.; Gimeno-Adelantado, J.V.; Vaiedelich, S. Gas chromatography/mass spectrometry characterization of historical varnishes of ancient Italian lutes and violin. *Anal. Chim. Acta* **2007**, *584*, 172–180. [CrossRef]
7. Echard, J.P.; Bertrand, L.; Von Bohlen, A.; Le Hô, A.S.; Paris, C.; Bellot-Gurlet, L.; Soulier, B.; Lattuati-Derieux, A.; Thao, S.; Robinet, L.; et al. The Nature of the extraordinary finish of Stradivari's instruments. *Angew. Chemie-Int. Ed.* **2010**, *49*, 197–201. [CrossRef]
8. Nagyvary, J. The chemistry of a stradivarius. *Chem. Eng. News* **1988**, *66*, 24–31. [CrossRef]
9. Echard, J.P.; Lavédrine, B. Review on the characterisation of ancient stringed musical instruments varnishes and implementation of an analytical strategy. *J. Cult. Herit.* **2008**, *9*, 420–429. [CrossRef]
10. Rovetta, T.; Invernizzi, C.; Licchelli, M.; Cacciatori, F.; Malagodi, M. The elemental composition of Stradivari's musical instruments: New results through non-invasive EDXRF analysis. *X-ray Spectrom.* **2018**, *47*, 159–170. [CrossRef]
11. Rovetta, T.; Invernizzi, C.; Fiocco, G.; Albano, M.; Licchelli, M.; Gulmini, M.; Alf, G.; Fabbri, D.; Rombolà, A.G.; Malagodi, M. The case of Antonio Stradivari 1718 ex-San Lorenzo violin: History, restorations and conservation perspectives. *J. Archaeol. Sci. Rep.* **2019**, *23*, 443–450. [CrossRef]
12. Zumbühl, S.; Soulier, B.; Zindel, C. Varnish technology during the 16th-18th century: The use of pumice and bone ash as solid driers. *J. Cult. Herit.* **2021**, *47*, 56–68. [CrossRef]
13. Invernizzi, C.; Fichera, G.V.; Licchelli, M.; Malagodi, M. A non-invasive stratigraphic study by reflection FT-IR spectroscopy and UV-induced fluorescence technique: The case of historical violins. *Microchem. J.* **2018**, *138*, 273–281. [CrossRef]

14. Su, C.; Chen, S.; Chung, J.; Li, G.; Brandmair, B.; Huthwelker, T.; Fulton, J.L.; Borca, C.N.; Huang, S.; Nagyvary, J.; et al. Materials Engineering of Violin Soundboards by Stradivari and Guarneri. *Angew. Chemie Int. Ed.* **2021**, *60*, 19144–19154. [CrossRef]

15. Obataya, E. Effects of natural and artificial ageing on the physical and acoustic properties of wood in musical instruments. *J. Cult. Herit.* **2017**, *27*, S63–S69. [CrossRef]

16. Tai, H.-C.; Chen, P.-L.; Xu, J.-W.; Chen, S.-Y. Two-photon fluorescence and second harmonic generation hyperspectral imaging of old and modern spruce woods. *Opt. Express* **2020**, *28*, 38831–38841. [CrossRef] [PubMed]

17. Cacciatori, F.; Sheets, A. *Reunion in Cremona. Tesori dal "National Music Museum" Vermillion, South-Dakota al Museo del Violino*; Catalogue edited by Fondazione Museo del Violino; Antonio Stradivari Cremona: Cremona, Italy, 2019.

18. Torrisi, F. *Il 'Mandolino Coristo' di Antonio Stradivari La sua Rinascita a Cremona Nell'anno 2000*; Cremona, Italy, 2002.

19. Rovetta, T.; Canevari, C.; Festa, L.; Licchelli, M.; Prati, S.; Malagodi, M. The golden age of the Neapolitan lutherie (1750–1800): New insights on the varnishes and decorations of ten historic mandolins. *Appl. Phys. A Mater. Sci. Process.* **2015**, *118*, 7–16. [CrossRef]

20. De la Rie, E.R. Fluorescence of Paint and Varnish Layers (Part I). *Stud. Conserv.* **1982**, *27*, 1–7.

21. De la Rie, E.R. Fluorescence of paint and varnish layers (Part II). *Stud. Conserv.* **1982**, *27*, 65–69.

22. Bonaduce, I.; Ribechini, E.; Modugno, F.; Colombini, M.P. *Analytical Chemistry for Cultural Heritage*; Springer International Publishing: Cham, Switzerland, 2017.

23. Artioli, G. *Scientific Methods and Cultural Heritage: An Introduction to the Application of Materials Science to Archaeometry and Conservation Science*; Oxford University Press: Oxford, UK, 2010.

24. Korte, E.H.; Staat, H. Infrared reflection studies of historical varnishes. *Fresenius. J. Anal. Chem.* **1993**, *347*, 454–457. [CrossRef]

25. Rosi, F.; Cartechini, L.; Sali, D.; Miliani, C. Recent trends in the application of fourier transform infrared (FT-IR) spectroscopy in Heritage Science: From micro: From non-invasive FT-IR. *Phys. Sci. Rev.* **2019**, *4*, 1–19. [CrossRef]

26. Dondi, P.; Lombardi, L.; Rocca, I.; Malagodi, M.; Licchelli, M. Multimodal workflow for the creation of interactive presentationsof 360 spin images of historical violins. *Multimed. Tools Appl.* **2018**, *77*, 28309–28332. [CrossRef]

27. Dondi, P.; Lombardi, L.; Invernizzi, C.; Rovetta, T.; Malagodi, M.; Licchelli, M. Automatic Analysis of UV-Induced FluorescenceImagery of Historical Violins. *ACM J. Comput. Cult. Herit.* **2017**, *10*, 1–13. [CrossRef]

28. Fiocco, G.; Gonzalez, S.; Invernizzi, C.; Rovetta, T.; Albano, M.; Dondi, P.; Licchelli, M.; Antonacci, F.; Malagodi, M. Compositional and morphological comparison among three coeval violins made by giuseppe guarneri 'del Gesù' in 1734. *Coatings* **2021**, *11*, 884. [CrossRef]

29. Beckhoff, B.; Kanngießer, B.; Langhoff, N.; Wedell, R.; Wolff, H. *Handbook of Practical X-ray Fluorescence Analysis*; Springer: Berlin/Heidelberg, Germany, 2006.

30. Poli, T.; Chiantore, O.; Nervo, M.; Piccirillo, A. Mid-IR fiber-optic reflectance spectroscopy for identifying the finish on wooden furniture. *Anal. Bioanal. Chem.* **2011**, *400*, 161–1171. [CrossRef] [PubMed]

31. Invernizzi, C.; Daveri, A.; Vagnini, M.; Malagodi, M. Non-invasive identification of organic materials in historical stringed musical instruments by reflection infrared spectroscopy: A methodological approach. *Anal. Bioanal. Chem.* **2017**, *409*, 3281–3288. [CrossRef]

32. Derrick, M.; Stulik, D.; Landry, J. *Infrared Spectroscopy in Conservation Science*; Getty Publications: Los Angeles, CA, USA, 1999.

33. Invernizzi, C.; Rovetta, T.; Licchelli, M.; Malagodi, M. Mid and near-infrared reflection spectral database of natural organic materials in the cultural heritage field. *Int. J. Anal. Chem.* **2018**, *7823248*, 1–16. [CrossRef]

34. Miliani, C.; Rosi, F.; Daveri, A.; Brunetti, B.G. Reflection infrared spectroscopy for the non-invasive in situ study of artists' pigments. *Appl. Phys. A Mater. Sci. Process.* **2012**, *106*, 295–307. [CrossRef]

35. Liu, Z.; Cao, J. Fabrication of superhydrophobic wood surface with a silica/silicone oil complex emulsion. *Wood Res.* **2018**, *63*, 353–364.

36. Fiocco, G.; Rovetta, T.; Gulmini, M.; Piccirillo, A.; Canevari, C.; Licchelli, M.; Malagodi, M. Approaches for detecting madder lake in multi-layered coating systems of historical bowed string instruments. *Coatings* **2018**, *8*, 171. [CrossRef]

37. Weththimuni, M.L.; Canevari, C.; Legnani, A.; Licchelli, M.; Malagodi, M.; Ricca, M.; Zeffiro, A. Experimental characterization of oil-colophony varnishes: A preliminary study. *Int. J. Conserv. Sci.* **2016**, *7*, 813–826.

38. Daher, C.; Paris, C.; Le Hô, A.S.; Bellot-Gurlet, L.; Échard, J.P. A joint use of Raman and infrared spectroscopies for the identification of natural organic media used in ancient varnishes. *J. Raman Spectrosc.* **2010**, *41*, 1494–1499. [CrossRef]

39. Daher, C.; Pimenta, V.; Bellot-Gurlet, L. Towards a non-invasive quantitative analysis of the organic components in museum objects varnishes by vibrational spectroscopies: Methodological approach. *Talanta* **2014**, *129*, 336–345. [CrossRef] [PubMed]

40. Mills, J.S.; White, R. Natural resins of art and archaeology their sources, chemistry, and identification. *Stud. Conserv.* **1977**, *22*, 12–31.

41. Tai, B.H. Stradivari's Varnish A Review of Scientific Findings—Part I. *J. Violin Soc. Am. VSA Pap.* **2007**, *21*, 119–144.

42. Von Bohlen, A.; Meyer, F. Microanalysis of old violin varnishes by total-reflection X-ray fluorescence. *Spectrochim. Acta-Part B At. Spectrosc.* **1997**, *52*, 1053–1056. [CrossRef]

43. Echard, J.P. In situ multi-element analyses by energy-dispersive X-ray fluorescence on varnishes of historical violins. *Spectrochim. Acta-Part B At. Spectrosc.* **2004**, *59*, 1663–1667. [CrossRef]

44. Malagodi, M.; Canevari, C.; Bonizzoni, L.; Galli, A.; Maspero, F.; Martini, M. A multi-technique chemical characterization of a Stradivari decorated violin top plate. *Appl. Phys. A Mater. Sci. Process.* **2013**, *112*, 225–234. [CrossRef]

45. Eastaugh, N.; Walsh, V.; Chaplin, T.; Siddall, R. *Pigment Compendium A Disctionary and Optical Microscopy of Historical Pigments*; Elsevier: Amsterdam, The Netherlands, 2013.
46. Canevari, C.; Delorenzi, M.; Invernizzi, C.; Licchelli, M.; Malagodi, M.; Rovetta, T.; Weththimuni, M. Chemical characterization of wood samples colored with iron inks: Insights into the ancient techniques of wood coloring. *Wood Sci. Technol.* **2016**, *50*, 1057–1070. [CrossRef]
47. Bonizzoni, L.; Canevari, C.; Galli, A.; Gargano, M.; Ludwig, N.; Malagodi, M.; Rovetta, T. A multidisciplinary materials characterization of a Joannes Marcus viol (16th century). *Herit. Sci.* **2014**, *2*, 15. [CrossRef]
48. Bulathsinghala, A.T.; Shaw, I.C. The toxic chemistry of methyl bromide. *Hum. Exp. Toxicol.* **2013**, *33*, 81–91. [CrossRef] [PubMed]

Article

Grouping Ceramic Variability with pXRF for Pottery Trade and Trends in Early Medieval Southern Tuscany. Preliminary Results from the Vetricella Case Study (Grosseto, Italy)

Cristina Fornacelli [1,*], Vanessa Volpi [2,3], Elisabetta Ponta [2], Luisa Russo [2], Arianna Briano [2], Alessandro Donati [3], Marco Giamello [1] and Giovanna Bianchi [2]

[1] Dipartimento di Scienze Fisiche, della Terra e dell'Ambiente, Università degli Studi di Siena, Via Laterina 8, 53100 Siena, Italy; marco.giamello@unisi.it
[2] Dipartimento di Scienze Storiche e Dei Beni Culturali, Università degli Studi di Siena, Palazzo San Galgano, Via Roma 47, 53100 Siena, Italy; vanessa.volpi@unisi.it (V.V.); elisabettaponta@gmail.com (E.P.); luisarus87@gmail.com (L.R.); arianna_briano@yahoo.it (A.B.); giovanna.bianchi@unisi.it (G.B.)
[3] Dipartimento di Biotecnologie, Chimica e Farmacia, Università degli Studi di Siena, Via Aldo Moro 2, 53100 Siena, Italy; alessandro.donati@unisi.it
* Correspondence: fornacelli@unisi.it

Citation: Fornacelli, C.; Volpi, V.; Ponta, E.; Russo, L.; Briano, A.; Donati, A.; Giamello, M.; Bianchi, G. Grouping Ceramic Variability with pXRF for Pottery Trade and Trends in Early Medieval Southern Tuscany. Preliminary Results from the Vetricella Case Study (Grosseto, Italy). *Appl. Sci.* **2021**, *11*, 11859. https://doi.org/10.3390/app112411859

Academic Editors: Anna Galli and Letizia Bonizzoni

Received: 17 November 2021
Accepted: 8 December 2021
Published: 14 December 2021

Publisher's Note: MDPI stays neutral with regard to jurisdictional claims in published maps and institutional affiliations.

Abstract: The characterization of archaeological ceramics according to their chemical composition provides essential information about the production and distribution of specific pottery wares. If a correlation between compositional patterns and local production centers is assumed, pottery manufacturing and trade and, more generally, economic, political, as well as cultural relations between communities and regions can be investigated. In the present paper, the combined application of portable XRF and statistical analysis to the investigation of a large repertory of ceramic fragments allowed us to group the assemblage by identifying geochemical clusters. The results from the chemical and statistical analysis were then compared with reference ceramic samples from the same area, as well as with macroscopic and petrographic observations to confirm, coalesce or subdivide putative sub-divisions. The study of 141 samples from different sites located within a wide area spanning across the Colline Metallifere and the coast (Monterotondo Marittimo, Roccastrada, Donoratico, and Vetricella) provided new clues for a new interpretive archaeological framework that suggests that there was a well-defined organization of pottery manufacturing and circulation across southern Tuscany during the early medieval period.

Keywords: pXRF; PCA; pottery

1. Introduction

The characterization of archaeological ceramics according to their elemental composition provides essential information about the circulation and consumption of specific pottery wares in the territory [1].

In recent years, the application of pXRF analysis to archaeological issues has become attractive in many ways. It has fast acquisition times and low experimental costs, and facilitates the scanning of large repertories in a non-destructive manner. Handled and portable devices also promote the in situ analysis of a wide range of materials [2].

Despite the challenges and limitations of using portable XRF techniques with source ceramics, the use of statistical methods has helped develop a fundamental tool for discerning between different pottery samples [3]. Aside from the accuracy of both the measurement and data processing, the way the results are inferred relies on how precisely the geographic location of the source material can be defined [4]. When a precise definition of the geographic location of the source material is not possible, or the geographical range is represented by a wide and highly heterogeneous area, the analysis of the reference materials (both concerning clays and/or reference pottery samples) becomes essential for inferring possible ancient exchange routes [4,5]. Several studies have demonstrated that pXRF analyses are applicable to the provenance studies of ceramics when the geological variability of the region and/or the inclusion of reference materials are also considered [6].

In this way, pXRF produces compositions and, subsequently, clusters that can be compared with other information (such as shape, surface finish and functional typology, as well as the indications that can be obtained via traditional mineralogical and petrographic observations) to confirm, coalesce or sub-divide putative sub-divisions of the pottery assemblage under study.

In the last few years, the interdisciplinary, ERC-advanced nEU-Med project promoted an in-depth archaeometric examination of a wide repertory of different ceramic fragments and raw materials that have been obtained from the main clay sources identified in the Colline Metallifere area [7]. Within this framework, the present study aims to describe how the application of pXRF to the screening of a large repertory of ceramic fragments has supported the acquisition of new data that helps to delineate the production and circulation of pottery during the Early Middle Ages (8th–10th centuries) within the Colline Metallifere area (southern Tuscany).

The archaeological site of Vetricella is a peculiar case study, as the archaeological data collected so far (not only concerning the ceramic findings) indicate that this settlement was the center of a royal property active between the 8th and 11th centuries [8]. Thanks to the archaeological campaigns promoted by the nEU-Med project, Vetricella represents the first Italian context of a royal property that has been extensively excavated so far.

Portable XRF has been used here for a preliminary analytical survey of a wide repertory of 141 ceramic fragments from Vetricella and the nearby archeological sites of Monterotondo Marittimo, Roccastrada and Donoratico (Figure 1). The pXRF analysis is also supported by further statistical, mineralogical and petrographic analyses. The chemical investigation and the subsequent statistical grouping also aim to optimize sample selection for a further in-depth characterization of clusters and outliers via high-resolution methods. The variability of the ceramics has been determined through a new approach that is based on the unsupervised classification of chemical elements that results from handheld XRF measurements [9], albeit controlled by archaeological and typological classifications, as well as traditional mineralogical and petrographic analysis.

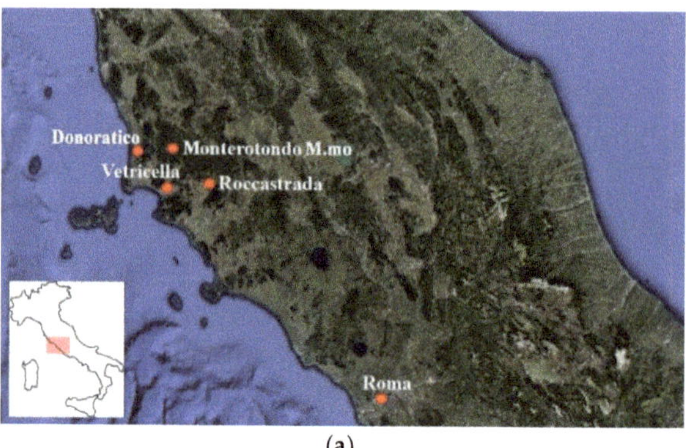

(**a**)

Figure 1. *Cont.*

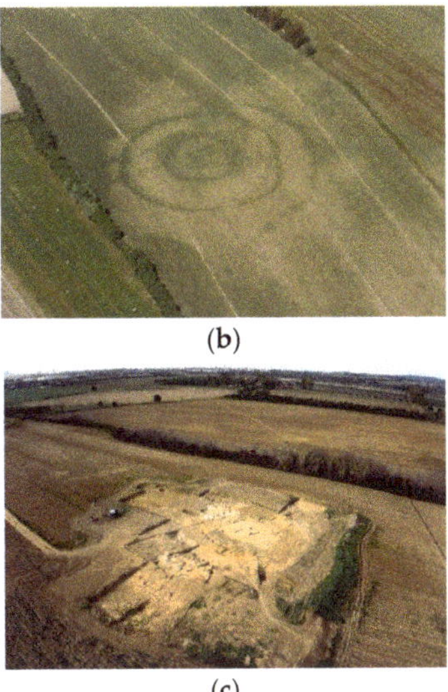

(b)

(c)

Figure 1. (**a**) Map of central Italy where the locations of the ceramic assemblages that have been explored in the text are reported. (**b**,**c**) The site of Vetricella.

2. Application of Portable XRF for Ceramic Investigations: The State of the Art

Highly sensitive benchtop analytical techniques for a chemical investigation of heterogenous samples, such as inductively coupled plasma mass spectrometry (ICP-MS), X-ray fluorescence (XRF), neutron activation analysis (NAA) and scanning electron microscopy and energy dispersive X-ray spectrometry (SEM-EDX), have been extensively used for the geochemical characterization of both ceramic materials and clay sources [10–13]. Despite their high accuracy and precision, these methods suffer from high costs and long acquisition times, together with the partial or total destruction of the sample [4].

In the last decades, the use of non-destructive methods in conservation science has gained general acceptance [14–18], however due to the reduced sensitivity to trace elements, the use of XRF in the investigation of archaeological materials remained poor compared to more accurate analytical techniques [19].

The successful application of pXRF to the investigation of highly heterogeneous materials, such as ceramics and soils, is, however, still strongly discussed [20–28]. Central to the debate is the complex nature of the matrix itself and the quest for both high analytical precision and reproducibility [29,30].

A number of dedicated analytical strategies have been proposed so far to overcome the main disadvantages that affect the application of pXRF to the investigation of ceramics. The analysis of homogenized powdered samples and the acquisition of multiple spots are methods used to average and assess the variations caused by the heterogeneity of the fabric (typically consisting either of a clay matrix and a mineral skeleton and/or a temper of variable particle size) and the possible presence of surface alterations (caused by firing or post-depositional conditions) or coatings [20,21,31].

A semi-quantitative estimation of the chemical concentration of each element could be achieved by the application of several calibration methods. Portable XRF spectrometers offer built-in calibration packages for a number of different substrates. Internal calibrations methods can also be adjusted by the user thanks to the measurement of certified reference materials (CRMs), or standards, which improves the accuracy of the reported concentrations within a specific dynamic range [23].

The application of built-in calibrations is, however, strongly matrix dependent, and the reduced versatility of internal software represents one of the main disadvantages to the use of pXRF in ceramic studies [19,20,23,32]. While pre-installed calibrations that are based on commercial algorithms are considered satisfactory for more homogenous substrates, the development of a material-specific calibration is strongly recommended to help users manage the complex nature of more heterogeneous materials. Previous studies, in particular, reported relative errors of around 50% for most factory calibrations, while custom standardizations typically induced a relative error of less than 10% [32–34].

The development of an empirical calibration for ceramic studies has thus been considered mandatory for obtaining valid and solid results for an inter-instrument comparison with data extracted by the tests, or for a future evaluation of a database outside the original experiment [21,32,35,36]. Nevertheless, despite the high reliability of the correction coefficients that is provided by empirical matrix-matched calibrations, the accuracy of an intra-instrument comparison is not without issues, since different instruments have been demonstrated to report varying values even following the same calibration method [37]. On the other hand, if the comparability of the dataset to others could not be assured, solid empirical evidence proves the adequacy of the application of built-in calibration routines for obtaining internally consistent result for archaeological interpretations [35,38].

The validation of dedicated internal calibrations in the analysis of ceramics represents an important challenge for making pXRF more accessible. For many users, a lack of access to more sensitive analytical techniques (e.g., ICP-OES, or NAA) for collecting calibration data for the creation of a matrix-matched calibration can be highly problematic and expensive, as can the acquisition of commercial calibration standards that are suitable for unprepared archaeological samples [24].

The recent advancements in hardware and software development for pXRF led to the release of more precise internal calibrations, and recent studies have called for an overall acceptance of specific built-in calibration algorithms for the semi-quantitative characterization of archaeological materials [3,24,28,29,38,39].

Robust multivariate statistical methods have been widely and successfully applied to the analysis of large chemical databases for provenance studies and for the recognition and characterization of local ceramic production to identify possible trading activities [4,25,40,41].

To better represent the high variability of a geochemical dataset, a pre-processing stage is thus required to standardize data and reduce distinctions in the magnitudes between major and trace elements [6,42,43]. While some recent studies have stressed the advantages of applying log ratio transformations to raw data [44–47], the use of standardized log10 data, combined with multivariate statistical methods, has been demonstrated to satisfactorily represent the fluctuations in the absolute concentrations of a geochemical dataset [48,49].

Qualitative pXRF analyses are frequently used in archaeology to determine the relative composition of a substance with the main aim of detecting distinguishable elements for further provenance studies [3,36–39]. A number of recent studies explored both the potentialities and limitations of the analysis of pottery using pXRF, which has thus been proven to be feasible tool for the investigation of individual assemblages of various ceramic wares [30,50].

One of the most frequently highlighted limitations of the use of pXRF in provenance studies is that is has a more restricted range of detectable elements than traditional analytical methods. Despite the fact that the sensitivity and number of elements that can be measured by pXRF devices is between 10 and 20, only a variable range of 5–10 elements are in general found to be useful. Moreover, the distinction between different ceramic groups is generally based upon elements that XRF methods are not great at detecting (e.g., rare earth elements) [6].

More sensitive analytical methods (e.g., INAA, ICP-MS, etc.) have been so far extensively used in provenance studies because they can be used as part of a high dimensional dataset (for INAA and ICP-MS, this is 23 and 19 elements, respectively), which comprehensively incorporates part of the spectrum of rare earth elements (REE), meaning that they can precisely segregate elemental groups on a microregional scale [1,51]. By contrast, most of these methods involve long analytical times, together with complex and destructive sample preparation. Even though the sample preparation issue could be solved by using laser ablation ICP-MS, its application to the analysis of heterogeneous materials is frequently semi-quantitative at best [35,52].

Despite the non-negligible discrepancies that are registered in the absolute elemental concentrations, non-destructive pXRF has been demonstrated to be able to distinguish between geochemically different ceramic pastes and identify groups in ways that closely correlate with those indicated by traditional methods [19–24]. In a recent paper [6], LeMoine provides important insights and suggestions, showing how pXRF and INAA produced comparative broad 'macro-provenance' groupings, with some ability of INAA to provide a more detailed sub-division of large groups.

Even though pXRF cannot substitute traditional analytical methods as a general elemental provenance-identifying procedure for ceramics, it can represent an extremely powerful alternative in cases where the internal grouping of 'closed' populations of delicate and/or highly abundant repertories is required [19]. The identification of carefully tailored measurement and processing protocols might provide good statistical results in accordance with those that are obtained through lab-based methods and can strongly support the archaeological interpretation [53].

The petrographic and microtextural investigation of selected fragments can help with identifying temper- and mineral-related variations that could influence the chemical database, as well as with evaluating various technological aspects to strengthen the interpretation of pXRF results [3,54]. The provenance may also be inferred from the mineralogical composition of the reference groups [13].

The characterization by pXRF of fragments whose provenance has been already determined by OM and/or traditional chemical analysis can support the definition of a database for the pXRF examination of unstudied samples from the same archives in collections where intrusive sampling is not allowed. On the other hand, the initial extensive pXRF screening of large ceramic repertories also promotes the preliminary definition of compositional clusters and outliers which can provide a selection of the most representative samples to be further investigated via more sensitive analytical methods. The in-depth characterization of selected samples, based on chemical grouping, can then help to refine the data processing by providing further indications about those elements which could properly describe the chemical dataset [5,55].

Within this framework, the present study aims to validate the great potential of pXRF to help with the investigation of a large repertory of pottery samples. In view of the results, the advantages of the application of a tailored analytical method to a set of reference samples are discussed together with some preliminary outcomes that have resulted from the analysis of samples which have an unknown provenance.

3. Archaeological Background

Previous archaeological studies on early medieval ceramic production in Tuscany have stressed the existence of a fragmented manufacturing base that consisted of only a few dispersed production sites until the 10th century. Within this framework, the archaeological evidence suggested that there was a local manufacture of cooking, table and storage wares, with a scattering of highly skilled ateliers being documented in some areas, including the Colline Metallifere [56–59].

The archaeological studies that have been carried out in the last few years within the framework of the ERC nEU-Med project [7] aimed to provide a more accurate chronological contextualization of the production of some of these ceramic classes between the 7th and 10th centuries. This also encouraged intense research activity that has been focused on the revision of past studies on some ceramic assemblages from different sites located within the Colline Metallifere [60–64]. In the wake of the recent advances in ceramic studies that have been promoted within the nEU-Med project, the more in-depth characterization of pottery manufacture in the Colline Metallifere area provided a solid point of comparison with the exceptional findings of Vetricella.

Vetricella and its surrounding territory formed an important royal property that preserved its status until the end of the 11th century. Since its discovery in 2005 during an aerial archaeology survey of the wide Maremma Grossetana coastal plain (southern Tuscany), the site of Vetricella immediately claimed the attention of scholars due to its many peculiarities. In particular, thanks to the exceptional variety of findings and its peculiar location in a territory that resembled many other coastal landscapes of the early medieval western Mediterranean, at a crucial junction between the sea and the Colline Metallifere, the site of Vetricella was then identified as key site within the framework of the nEU-Med project [8].

The extensive archaeological excavations carried out at Vetricella in the last years returned a variety of indicators (numismatic, ceramic and glassy materials, together with a large amount of archaeometallurgical, archaeozoological and archaeobotanical findings). The exceptional nature of these findings, combined with a detailed investigation of the historical documents, provided clues about the definition of the site as a royal property, characterized as it was by extensive economic and productive activity dating between the 9th and 11th centuries, while evidence of an earlier occupation (7th–8th centuries) was also observed [65].

Due to the variety of findings returned at Vetricella, the integration of the archaeological research with disciplines that include geoarchaeology, archaeobotany, archaeometallurgy, physical anthropology, and physical chemistry was first of all essential to reconstructing a general history of the landscape, as well as for defining the correlations between Vetricella and the exploitation of the widely heterogeneous natural resources of the Colline Metallifere [7,8].

The ceramic assemblage of Vetricella includes a large repertory of pottery belonging to different ceramic functional classes, such as cooking ware, tableware and storage products (jars, jugs, pots, bowls and small amphorae), as well as the highly skilled *Forum* and *sparse-glazed* wares. The significant amount of pottery dedicated to different uses and related to well-defined stratigraphical contexts represented an optimal repertory with which to deepen the investigation into pottery production in south-western Tuscany during the Early Middle Ages.

A preliminary comparison of the ceramic assemblage returned at Vetricella with those from other sites located within the Colline Metallifere (e.g., Monterotondo Marittimo and Roccastrada) and the coast (Donoratico) allowed us to define the existence of a complex and well-organized manufacturing base that was characterized by large-scale distribution within the territory (dating back to the 7th–8th centuries and with a recognizable increase from the 9th until the 11th centuries).

The multidisciplinary investigation then started to help us form an in-depth characterization of the ceramic assemblages as they related to different sites that are located within the broad and heterogenous Colline Metallifere area, which would in turn help us to identify possible production areas and distribution flows within the territory. The results from this study also provided new perspectives on the distribution of goods within Colline Metallifere during the 7th–11th centuries and suggested that Vetricella played a key role as a central site dedicated to the collection and re-distribution of a wide and heterogeneous range of locally produced materials.

4. Geological Background

The geological setting of the area is the result of a sedimentary and tectonic evolution that affected the Italian peninsula between the Oligocene and the Pliocene, and which was strongly connected to the several superimposed compressive stages that led to the formation of the nappes of the Northern Apennines [66]. The heterogenous structural and stratigraphic setting of southern Tuscany is therefore derived from different deformational processes that are related to the convergence between the Corsica–Sardinia and Apulia microplates [67] and to the following post-collisional extensional tectonics [68].

Between the Oligocene and lower Miocene, the compressional deformations related to the collisional episodes resulted in the development of a nappe sheet stack consisting of different paleogeographic domains.

From the late Miocene, post-collisional magmatism affected southern Tuscany and the northern Tyrrhenian Sea. The diffused acidic peraluminous intrusive bodies (Elba, Montecristo, Giglio, Gavorrano, Campiglia Marittima) and cogenetic volcanic products (rhyolites of San Vincenzo, Roccastrada, La Tolfa) originated from the intense igneous activity that today defines the so-called Tuscan Magmatic Province (TMP) [69]. The TMP consists of three main magmatic typologies (melts originating from crustal anatexis, melts resulting from the mixing of acid crustal and basic subcrustal melts, and mantle-derived melts) among which the Roccastrada rhyolite represents the most acidic term [70].

The emplacement of magmatic rocks and the related intense hydrothermal circulation led to the formation of several mineral deposits [71]. Polymetallic sulphidic deposits (Cu-Pb-Zn-Ag) of southern Tuscany are mainly hosted along a belt which extends from the Tyrrhenian coast towards the province of Siena.

Between the middle Miocene and the Pliocene, the occurrence of extensional tectonics and crustal thinning led to the development of sedimentary basins with significantly different tectonic and stratigraphic features [72]. The sediments filling the extensional basins formed the so-called Neo-autochthonous Complex [73], representing the uppermost post-orogenic sedimentary succession of southern Tuscany and mainly consisting of limestone breccias, fluvial conglomerates and calcareous tufa, together with lacustrine-marine (late Tortonian-Messinian), marine (Pliocene) and lacustrine-fluvial (early Villafranchian-Quaternary) clays [66,72,74] (Figure 2).

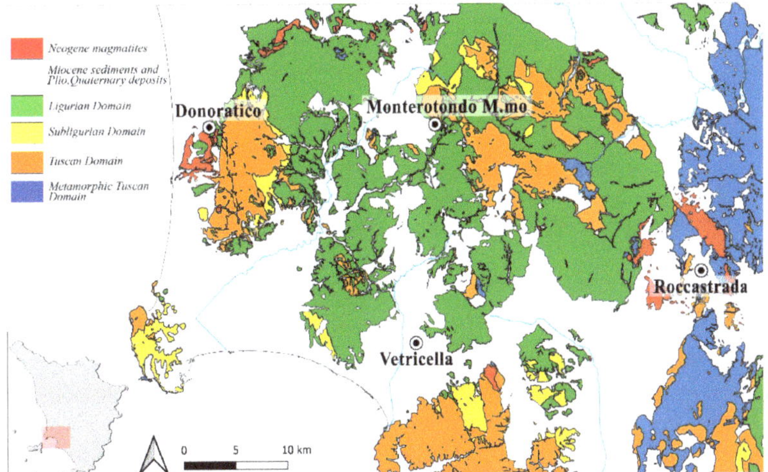

Figure 2. Geological map of southern Tuscany.

5. Materials and Methods

For the present study, the selected ceramic repertory from the Vetricella archaeological site includes 83 fragments (Supplementary Materials, Table S1), together with reference samples (Supplementary Materials, Table S2) consisting of 58 fragments from Monterotondo Marittimo (16), Roccastrada (13) and Donoratico (29).

The ceramic assemblage (Figure 3) consists of fine- to coarse-grained pottery for different uses (tableware, storage ware, cooking ware) which dates back to the 7th–10th centuries. Some of the samples are characterized by the presence of a Pb-based coating on the external surface (*sparse-glazed* ware).

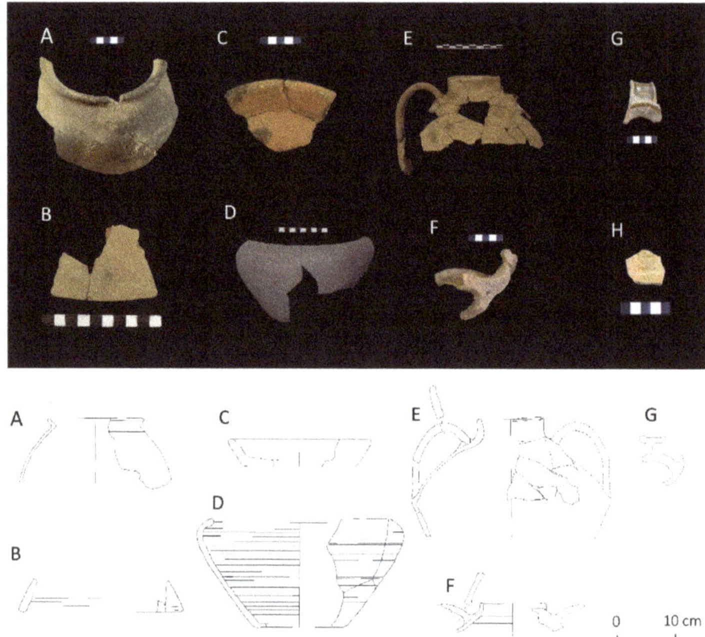

Figure 3. Some fragments from the investigated ceramic assemblage: (**A–C**) cooking ware; (**D**) table and storage ware; (**E,F**) small transport amphorae; (**G,H**) *sparse-glazed* ware.

5.1. Experimental

The geochemical investigation of the repertory was carried out via a combined pXRF and is supported by references to traditional mineralogical and petrographic studies.

The mineralogical and structural investigation was performed on thin sections using a LEICA DMRX optical polarizing microscope to provide a detailed characterization of both the matrix (homogeneity, color, iso-orientation and porosity) and the inclusions (composition, grain size, frequency, roundness, sorting) that are either naturally present or artificially added. The textural elements were estimated using suitable comparative charts [51,75,76].

The chemical characterization was performed using portable X-ray fluorescence (pXRF) with an Olympus INNOV-X Delta Premium DP-6000-C that was equipped with a 40 kV, 4 W and 200 microampere X-ray tube, Rh anode and a large-area SSD detector. All of the pottery samples were cleaned with water and an ordinary nylon-bristle toothbrush to remove the soil and particulate matter that was loosely adhered to the surface. Coatings were also removed when present. A representative portion of each fragment (proportional to the average crystal size) was then powdered to ensure the proper homogenization of the sample [77].

The analysis was performed in soil mode, which has proved in the past to be suitable for ceramic studies [78,79]. Even though major elements, such as Mg, Si or Al, cannot be determined, 'soil' mode analysis yields a large suite of minor and trace elements. The selected lifetime of the measurements, which were carried out in air, was 30 s for each sample, measuring in three energy ranges (two at 40 KeV and one at 15 KeV) for the analysis of the elements with the lowest atomic weight. During each measurement, a photograph of the analyzed area was recorded with the integrated camera. All of the samples were analyzed at three different spots in order to account for any uncertainties that were potentially introduced by the method or by the selection of the measurement area. In previous studies, up to five measurements have been suggested for the pXRF analysis of coarse ceramics in order to collect results with adequate precision [24]. A set of 33 elements were measured, but those with concentration values below or near the LOD were cleared from the list. Fourteen elements were at last used for an initial statistical evaluation: As, Ca, Cr, Cu, Fe, K, Mn, Ni, Pb, Rb, Sr, Ti, Zn and Zr.

The soil method is a Compton normalization approach that is calibrated for the analysis of elements whose chemical concentration is less than 2%, although good results can also be obtained for higher concentrations [80]. To ensure the highest reproducibility of the measurements, the instrument was used in benchtop mode and the samples were placed in special Teflon sample holders and sealed with a Mylar film.

5.2. Data Processing

In order to determine the most suitable set of elements for investigating the variance of the whole repertory of ceramic fragments, the relative standard deviations (RSD) were calculated (Table 1).

Table 1. The RSDs that are representative of each element, calculated from all assays taken from the powdered samples. RSD values of the whole repertory, as well as from the assemblages from the reference sites of Castellina, Donoratico, Monterotondo Marittimo and Roccastrada, are reported.

TOTAL													
Element	K	Ca	Ti	Cr	Mn	Fe	Ni	Cu	Zn	As	Zr	Pb	Rb	Sr
Mean (ppm)	26,868	26,662	4197	221	639	34,949	75	29	76	40	209	1781	186	163
Population STD	4240	28,373	815	141	356	10,049	42	13	20	67	59	4790	58	104
RSD %	16	93	19	64	56	29	56	46	26	167	28	269	31	64
Castellina													
Element	K	Ca	Ti	Cr	Mn	Fe	Ni	Cu	Zn	As	Zr	Pb	Rb	Sr
Mean (ppm)	28,360	29,768	4139	171	567	31,807	60	25	72	34	228	482	214	192
Population STD	4284	28,531	743	101	311	9820	31	10	19	30	63	2107	52	96
RSD %	15	89	18	59	55	31	52	40	26	87	27	437	24	50
Donoratico													
Element	K	Ca	Ti	Cr	Mn	Fe	Ni	Cu	Zn	As	Zr	Pb	Rb	Sr
Mean (ppm)	24,651	5588	4982	440	695	43,026	139	30	78	51	210	3916	128	69
Population STD	2089	1380	320	107	414	7874	23	10	15	110	33	5522	10	10
RSD %	8	25	6	24	59	18	16	34	20	216	15	141	8	14
Monterotondo Marittimo													
Element	K	Ca	Ti	Cr	Mn	Fe	Ni	Cu	Zn	As	Zr	Pb	Rb	Sr
Mean (ppm)	27,318	39,202	4339	178	924	39,742	70	35	90	30	147	24	146	171
Population STD	4529	20,773	687	23	336	5501	28	11	13	23	29	6	50	97
RSD %	17	53	16	13	36	14	40	30	15	75	20	24	34	57
Roccastrada													
Element	K	Ca	Ti	Cr	Mn	Fe	Ni	Cu	Zn	As	Zr	Pb	Rb	Sr
Mean (ppm)	22,435	8285	3120	152	510	29,533	41	26	67	35	159	39	201	82
Population STD	2728	2643	538	47	339	7915	13	8	17	44	43	10	47	38
RSD %	12	32	17	31	66	27	32	31	25	127	27	27	24	47

If the whole repertory of samples is considered, most of the elements showed an RSD% ranging between 16 and 93%, with the exception of As and Pb, which were characterized by higher values (167% and 269%, respectively). From a more in-depth investigation of the single repertories from the reference sites, the reasons for the high RSD values for As and Pb were found to have resulted from their being mainly correlated to the *sparse-glazed* ware samples from Donoratico, suggesting that an external factor was responsible for the observed Pb fluctuations. The *sparse-glazed* wares consisted of single-fired lead-glazed ceramics, where a suspension of PbO (alone or mixed with SiO_2) was applied to leather-hard bodies to obtain a transparent coating [81]. Depending on the firing conditions, a more or less extended diffusion of lead within the ceramic body could occur and the detection of high levels of lead in some of the glazed samples was considered to be a consequence of the incomplete removal of the external layers during sampling procedures.

The exploitation of a number of different lead deposits during the late Roman and early medieval periods has been widely documented [59,82–84]. Arsenic (together with other elements, such as silver, iron, or zinc) is frequently associated with poly-metallic sulfide deposits and could be present as an impurity in the final product [85].

The investigation of Pb and As levels in glazed ware samples showed a peculiar correlation between those fragments where the use of lead from southern Tuscany was assumed [10] and the presence of high levels of arsenic was again related to the diffusion of the glazing mixture within the ceramic body (Figure 4). The Roccastrada assemblage was also characterized by a high RSD% value for As. The reasons for the As anomaly were in this case related to the geology of the area and, in particular, to the intense hydrothermal activity that affected the primary rhyolitic rocks and which caused an enrichment in As [70,86].

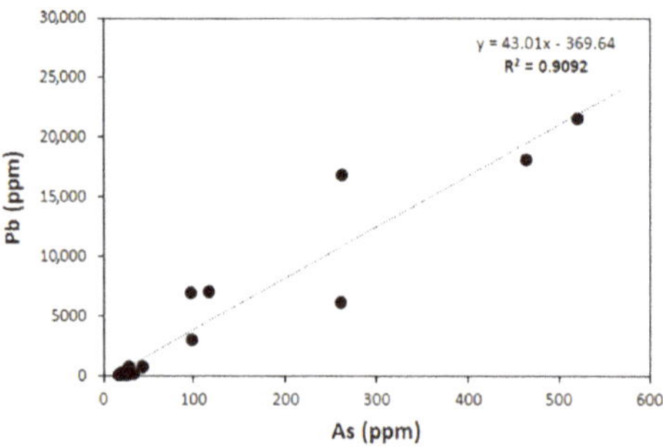

Figure 4. Pb/As bi-plot of glazed ware. Trend line and r^2 are reported.

Despite the fact that As could represent a marker that would help us discriminate between the use of raw materials from Roccastrada and other contexts, the anomaly that was registered for Donoratico suggested that lead and arsenic should be excluded from the analysis. Together with Pb and As, those elements that are known to be affected by other external factors (Cl, P, Ba and S) have also been excluded [21]. Due to the diffused presence of mixed sulfide formations in southern Tuscany, the inclusion of Cu and Zn in the dataset was considered to be significant for a more detailed grouping of the samples.

The chemical dataset was then pre-treated in order to minimize the influence of matrix effects and reduce distinctions in the magnitudes between major and trace elements [87]. Previous studies have discussed the identification of a more suitable type of transformation for investigating a closed archaeological dataset [42,44,47,48,88–90]. Among other techniques, the normalization of the data according to titanium and the standard log10 transformations have been compared in the present study.

Normalization according to a major element (or signals from the cathode) allowed us to limit experimental errors and fluctuations (e.g., irregular surfaces, small variations of the sample–instrument distance) and helped us ensure a higher level of analytical reproducibility [91,92]. Spearman's correlation was used to investigate the relationships between major, minor and trace elements, and titanium was then identified as the most representative element for the present chemical dataset [47,93].

On the other hand, log10 transformation was considered, as it compensates well for the differences in magnitudes between the major elements and the trace elements [42,87].

6. Results

The petrographic investigation of reference samples allowed us to identify the main mineralogical features that correspond to different areas of production (Table 1).

The fragments from Donoratico (fabric group DON, Figure 5a) were characterized by quartzose–feldspathic inclusions, which were the main mineralogical phases, with minor quantities of micas, opaque minerals, secondary calcite and rare pyroxenes. Lithic fragments were also observed, which mainly consisted of meta-arenites and siltites, together with rare intrusive rocks. Previous studies on the mineralogical and chemical features of these samples [10] also showed the diffuse presence of Fe and Ti oxides, together with lower amounts of spinel (mainly from the chromite series), titanite, monazite, and apatite.

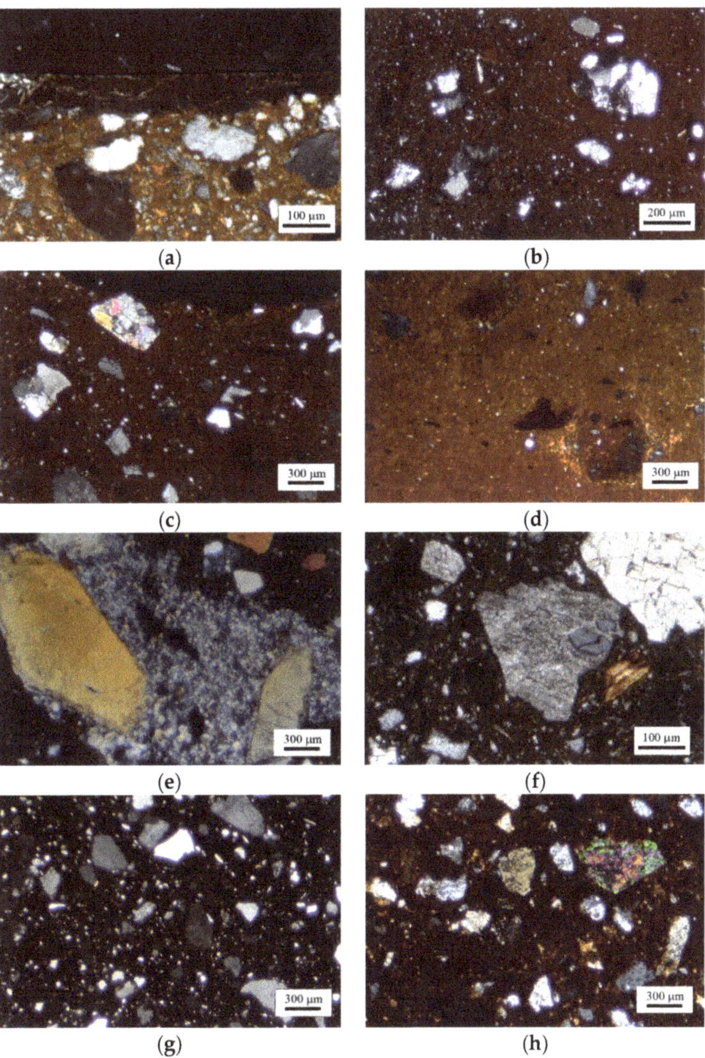

Figure 5. Microphotograph of selected samples from the fabric groups identified following the petrographic investigation: (**a**) sample DON194a, fabric group DON; (**b,c**) samples RA22 (fabric group MR1) and MR21 (fabric group MR2) from Monterotondo Marittimo; (**d**) sample RS35, fabric group RS1; (**e**) sample RS87, fabric group RS2; (**f**) sample RS24, fabric group RS3; (**g**) sample CSN124, fabric group RS4; (**h**) sample CSN221, fabric group EXT.

The ceramic assemblage from Monterotondo Marittimo was characterized by its having calcareous pastes where the predominant mineralogical inclusions were represented by quartz, together with feldspars (both plagioclase, K-feldspar), biotite, secondary calcite granules, sub-rounded rock fragments quartz arenite, siltstones and siliceous limestones. Minor amounts of clay pellets, opaque minerals and relics of microfossils (mainly ostracods) were also observed (fabric group MR1, Figure 5b), while rare mafic intrusive rock inclusions were detected in some of the samples (fabric group MR2, Figure 5c).

The fabrics from Roccastrada were characterized by the use of high-silica mineral inclusions, which were mainly represented by mono-crystalline quartz grains with sub-angular shape and a high degree of particle size standardization, suggesting a deliberate addition of a siliceous sand to the clay (fabric groups RS1, RS2, RS3 and RS4, Figure 5d–g, respectively). Subordinate amounts of rock fragments (poly-crystalline quartz and quartzite) and inclusions of volcanic origin (sub-rounded fragments of rhyolite and/or glass) were also observed, together with feldspars (mainly plagioclase), biotite, clay pellets, rare pyroxenes and opaque minerals.

Petrographic observations of the Vetricella assemblage allowed us to assign most of the samples to the above-described fabric groups. Only a few glazed ware samples showed peculiar features, suggesting the different provenance of the raw materials (likely extra-regional). A further fabric group was then identified (fabric group EXT, Figure 5h). All of the samples belonging to EXT group were characterized by large amounts of calcareous/dolomitic pastes where the predominant mineralogical phases were represented by quartz (mono and polycrystalline) and feldspars, together with frequent carbonates and micas. The carbonates mainly consisted of variable amounts of spathic and/or micritic calcite, together with rare bioclastic and/or dolomitic inclusions. Subordinate quantities of opaque minerals, rock fragments (flint and rare sandstone with clayey matrix) and rare pyroxenes and garnet inclusions were also observed.

Principal component analysis of the Ti-normalized and Log10 datasets was first of all performed on reference samples. Four components were extracted that represented 79.23% and 80.39% of the total variance of Ti-normalized and Log10 data, respectively. Biplots of the two first components of each PCA are shown in Figure 6a,b.

For the Ti-normalized dataset, PC1 (principal component 1) represented 38.90% of the total variance and allowed us to distinguish between high-Ca and low-Ca pastes, as well as between more weathered items (such as Rb-rich kaolinized intrusives, or shales). On the other hand, PC2 represented 15.24% of the total variance and was influenced by the presence of Cr- and Ni-bearing mineral phases.

The log10 statistical dataset, where PC1 and PC2 described 37.47% and 24.42% of the total variance, respectively, provided a better grouping of the samples. Contrary to Ti-normalized dataset, PC1 and PC2 are herein described by the nature of the mineral inclusions and the nature of the clay, respectively. From both the Ti-normalized and log10 PCAs it was possible to easily distinguish between the pastes from Donoratico from those of the internal areas (Monterotondo Marittimo and Roccastrada), where the main discriminant in represented by the use of a different temper in association with more or less calcareous clays. Due to the calcareous/dolomitic nature of the clays that were used for glazed ware pastes, the samples belonging to the EXT group represented the more calcareous terms of the series, but they could not be discriminated from local high-Ca pastes.

Once the main features of the reference samples were defined, the statistical analysis was then extended to the whole repertory to investigate the chemical variance of the samples from Vetricella. Four components were extracted that represented 79.91% and 80.19% of the total variance of Ti-normalized and Log10 data, respectively. Biplots of the two first components of each PCA are shown in Figure 6c,d. PC1 and PC2 represented, respectively, 32.35–29.91% and 36.02–20.74% of the total variance for Ti-normalized and log10 PCAs, respectively.

The PCA analysis of the whole repertory showed the highly heterogenous nature of the ceramic assemblage from Vetricella, with data superimposed on the groups that were identified by the reference samples. The petrographic analysis of the samples also provided a better assessment of the variations due to temper-related issues.

The combined use of PCA and petrographic analysis indicated a large incidence of Roccastrada pastes among the Vetricella fabrics (67%), while Monterotondo Marittimo (19%), Donoratico (10%) and extra-regional (4%) pastes were less represented.

The incidence of the different ceramic functional classes or shapes within the Vetricella assemblage is reported in Figure 7a. If the results from the petrographic and chemical analysis are considered, the samples belonging to the Roccastrada ateliers consisted of small amphorae (55%), likely used for the transportation of goods (so-called small transport amphorae, or STA), cooking ware (23%) and table and storage ware (21%). On the contrary, the samples with a geochemical and mineralogical association with the Monterotondo Marittimo area showed a higher incidence of table and storage ware (50%), together with minor amounts of STAs (38%) and cooking ware (13%). Finally, glazed ware represented the only ceramic class representative of the Donoratico region and extra-regional production (Figure 7b).

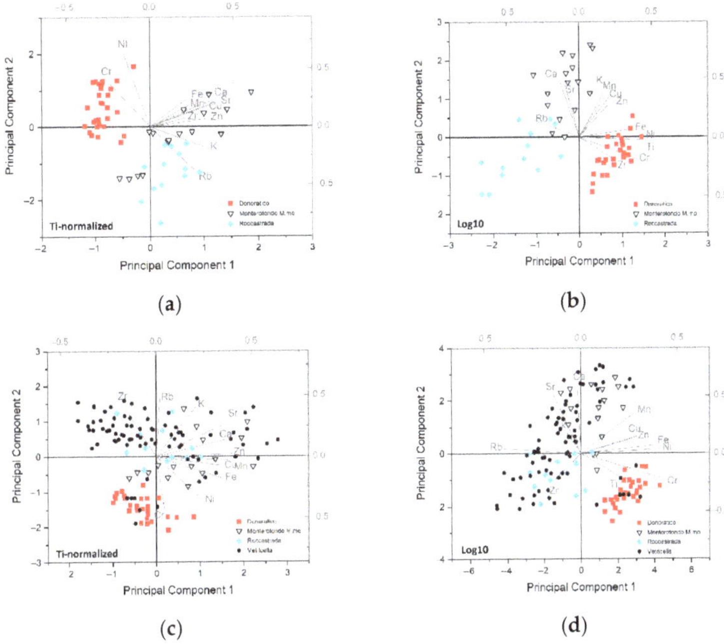

Figure 6. Bivariate plots from the PCA analysis of the Ti-normalized and log10 datasets relative to: (**a**,**b**) the reference ceramic samples; and (**c**,**d**) the whole repertory.

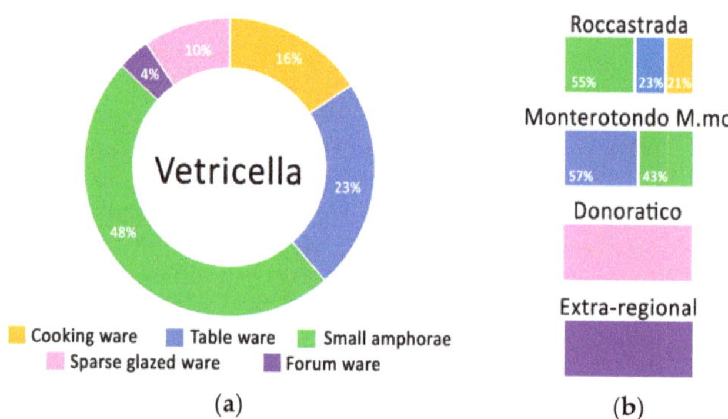

Figure 7. Outline of the main features of the Vetricella assemblage investigated in the present study according to the results from the archaeological, geochemical and petrographic analyses: (**a**) pie chart showing the main ceramic functional classes; (**b**) tree charts showing where the Vetricella assemblage is associated with different production areas depending on the mineralogical and geochemical nature of the pastes. The incidence of each ceramic functional class is also reported.

7. Discussion

The results of the investigation of the reference samples indicated how assemblages and materials can be distinguished from each other through the combined use of petrographic and non-destructive pXRF analysis. Based on the elements used in this analysis, the geochemical signatures of the samples showed more or less pronounced differences between the ceramic assemblages originating from different sites, while the petrographic investigation of the ceramic pastes provided a more detailed contextualization of the geochemical data.

Both mineralogical and geochemical data were strongly in agreement with the complex geology of southern Tuscany.

The reference samples from Roccastrada mainly consisted of cooking ware where the intentional addition of a temper, represented by abundant quartz with subordinate and variable amounts of clay pellets, rhyolite and volcanic glass, was observed. In the PCA bi-plots, the samples from Roccastrada formed a more dispersed group that was characterized by a good correlation with Rb, where the abundance and the mineralogical composition of the temper played a significant role in the definition of slight discrepancies in the geochemical signature.

The geology of the Roccastrada area is characterized by a diffused presence of Miocenic marine clays, represented by fine-grained blue clays with quite abundant micro-fauna (*Argille Azzurre* Formation) and gray clays with intercalated fine-grained sandstones (*Torrente Raquese* Formation, or *Pycnodonta* clays). The abundance of refractory materials, such as volcanic rocks (rhyolite and volcanic glass) and highly siliceous rock assemblages, promoted an intense production of refractory ceramic, such as cooking ware and technical ceramic for smelting activities.

The rhyolites of Roccastrada mainly consist of glass (53%), quartz (14–15%), K-feldspar 16%, plagioclase (8–9%), biotite (3–6%) and cordierite (2%), with minor apatite and zircon [70]. Rb^+ substitutes frequently for K^+ in K-feldspars, micas and clay minerals, with a natural tendency to concentrate during the weathering of the parent rock due to ion exchange and adsorption mechanisms. In magmatic systems, Rb^+ highly concentrates during late stages and could substitute for K^+ in K-bearing phyllosilicates (such as mica and biotite) and, to a lesser extent, in K-feldspar [86]. During the first weathering stages, the dismantling of biotite and the plagioclase present in rhyolitic rocks leads to a depletion in Rb. In the last stages, following the decomposition of K-feldspars and the alteration of zircon, a consequent increase in Rb depletion, together with K and Zr, is documented [70].

As observed for the samples of Roccastrada, the heterogeneous textural and mineralogical features observed for the fragments from Monterotondo Marittimo represented the main reason for higher fluctuations in the chemical composition. Most of the fine- and medium-grained vessels from this group were mainly represented by a good correlation with Ca (together with Sr) and K, while some coarse ware fragments constituted an intermediate cluster that was represented by Rb.

The correlation of fine- and medium-grained samples with Ca was in accord with the geology of Monterotondo Marittimo, as it is mainly represented by the diffuse presence of Miocenic blue marine clays, together with limestones, marl and polygenic marine conglomerates. Quartzose-feldspathic-micaceous sandstone (Macigno Formation) are also abundant, while sporadic gabbroic intrusions have been observed within the more extended outcrops related to *Palombino* shales. The use of a quartzose-feldspathic temper and/or a less purified clay, or at least a different clay source, was assumed to be responsible of the peculiar association of the coarser fabrics with Rb.

As for the fine-grained ware from Monterotondo, the glazed ware fragments belonging to the EXT group were well represented by Ca. The chemical and mineralogical composition of this group accords well with the geology of the Roman and northern Latium areas, characterized by the presence of Pliocenic marls and basic volcanites [94,95].

Finally, the *sparse-glazed* ware from Donoratico represented a well-defined group characterized by a peculiar correlation with Cr, Ni and Ti. Highly serpentinized ophiolite fragments, in addition to the presence of chromite and abundant quartz, together with minor amounts of Cr-spinel and Cr/Ni-rich pyroxene, have been documented in the sedimentary materials from Val di Cecina [96]. The identification of Cr- and Ni-bearing minerals was closely associated with the fractioning processes that are produced by the weathering and dismantling of the ophiolitic outcrops present in the area.

Previous studies on *sparse-glazed* ware manufacture in Donoratico confirmed the existence of dedicated ateliers in the area and the exploitation of local raw materials [59].

The introduction of the fragments from Vetricella in the statistical database revealed a wide distribution within the compositional clusters that was defined by the reference samples and which suggested the existence of a wide circulation of pottery converging towards the southern Tuscany coast.

The results from the PCA were in agreement with the petrographic analysis of the fabrics and paved the way for important considerations about the circulation and consumption of ceramics in the area during the Early Middle Ages that could be supported by further archaeological and archaeometric investigation on a large scale.

The first indications arising from the results of the present study stressed the central role of Vetricella as a collector of goods from different areas of southern Tuscany and provided impetus for future investigations that might help formulate a more detailed definition of the trading routes to and from Vetricella (Figure 8).

The comparison of the results of the geochemical and petrographic examinations of the studied repertory suggested a well-defined exploitation of raw materials across the Colline Metallifere in relation to the manufacture of artefacts with different intended uses. From the analysis of the whole repertory, it was possible to observe how coarse-grained pastes mainly showed a higher incidence of geochemical and mineralogical markers for Roccastrada, while fine- and medium-grained fabrics were equally distributed among the Monterotondo and Roccastrada fabric groups, as well as within the geochemical clusters. The *sparse-glazed* ware from Vetricella finally showed strong analogies with the Donoratico assemblage.

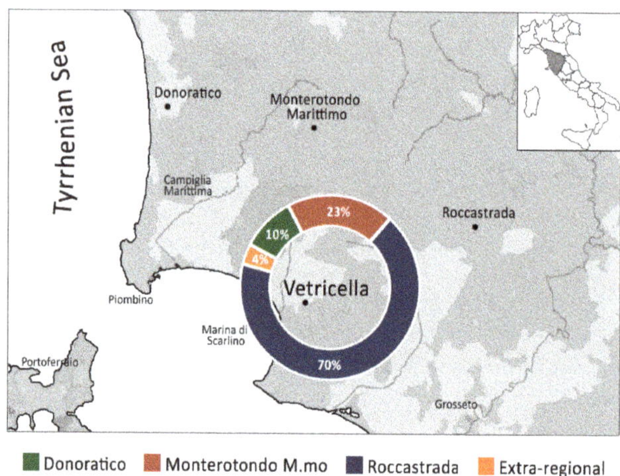

Figure 8. Outline of the results from the geochemical investigation of the ceramic assemblage from Vetricella, overlaying the geographical maps of the area. The pie chart represents the Vetricella assemblage according to the provenance of the pastes from different areas.

Small amphorae (STA) represent an interesting case study. This peculiar, double-handled closed shape was first identified during previous archaeological investigations in different sites in southern Tuscany [7]. As it stands at present, it is not possible to provide an exhaustive contextualization of the use and circulation of STAs in the region. Nevertheless, the exceptional number of STA fragments returned from the last excavations in Vetricella provides new clues to support further research within this framework.

In the present study, STAs represented 48%, 55% and 38% of the whole ceramic assemblage from Vetricella, Roccastrada and Monterotondo Marittimo, respectively. STA fragments from Vetricella, in particular, had a highly heterogenous composition where the chemical and mineralogical markers of both Monterotondo Marittimo (23%) and Roccastrada (77%) were observed. Previous studies also showed the presence in Vetricella of a few fragments related to some ateliers located in the coastal area [61].

The relevant amount of STAs produced with local raw materials suggested the extensive manufacture of handled closed shapes, likely intended for circulation as a local alternative to the import of amphorae from the south of the peninsula, though this traffic is still scarcely documented along the coastal and sub-coastal strip of Tuscany [97], and evidence regarding the nature of the transported materials are still poor.

Further investigations will be thus vital to provide an accurate characterization of STA manufacture and consumption, as well as a detailed definition of the exchange routes within the territory.

8. Conclusions

The present work aimed to contribute to the long-lasting debate surrounding the successful use of pXRF with archaeological materials. While the widespread application of portable analytical methods to archaeological issues prompted a number of multidisciplinary studies, the non-negligible limits in the investigation of highly heterogenous materials restricted the use of pXRF to pottery assemblages.

Based upon the indications arising from the comparison with the more solid petrographic analysis and the application of statistical methods, pXRF was here used as a powerful tool for a broad 'screening' of a large repertory of fragments and provided a reliable internal grouping of a 'closed' population of samples.

In the future, the limitations arising from the detection of chemical elements within a restricted dataset could be reduced via the application of more sensitive analytical methods on selected samples. The detection of a wider array of chemical elements (comprehensive of light elements and REE) will help to more accurately distinguish between the sub-groups, and lead to the creation of a well-defined database to support further pXRF examinations of additional unstudied fragments.

The combined use of pXRF and statistical analysis facilitated the geochemical grouping of the ceramic assemblage, providing clues about the convergence of different commercial routes towards the site of Vetricella that reveal its key role as a central site dedicated to the collection and re-distribution of a wide and heterogeneous range of goods that were produced in the area, as also indicated by the presence of glazed pottery that required highly skilled labour.

The petrographic investigation also supported a more in-depth analysis of the fabrics, particularly concerning those samples where the chemical clustering was uncertain.

Despite the fact that the repertory that we studied here represents just a limited selection of samples from all of the sites cited in the present work, the results allowed us to show how ceramic manufacture was diffused via trading capillaries across a wide area, i.e., the Colline Metallifere. In contrast with previous studies, the study also suggested that the diffused ateliers were tightly interconnected, while the archaeological data indicated how the ceramic production was strongly dependent on the complex management of local resources that were under the control of the royal authority.

In the debate about pXRF application in the archaeological sciences, the recent huge increase in the use pXRF instruments by conservators and archaeologists represents another controversial point [19]. Beside the purely analytical and technical issues, the present work also wanted to stress the importance of tight cooperation between scientists and archaeologists, which would help us to develop a more solid analytical strategy and provide reliable answers to archaeological and historical questions.

The interdisciplinary interpretation of the analytical data here also paves the way for the organization of future investigations that are focused on different and more specific topics where the application of more solid analytical methods could help to delineate a more accurate picture of pottery circulation to and from Vetricella on a larger scale.

Supplementary Materials: The following are available online at https://www.mdpi.com/2076-3417/11/24/11859/s1, Table S1: Summary of the main petrographic features of the ceramic assemblage from Vetricella. Samples have been grouped after the identification of the area of provenance of the raw materials according to the mineralogical composition of the ceramic bodies. Table S2: Summary of the main petrographic features of the reference ceramic assemblages from Monterotondo Marittimo, Roccastrada and Donoratico. Samples have been grouped after the identification of the area of provenance of the raw materials according to the mineralogical composition of the ceramic bodies.

Author Contributions: Conceptualization, C.F., V.V., E.P., L.R. and G.B.; Data curation, C.F.; Formal analysis, C.F. and V.V.; Funding acquisition, G.B.; Investigation, C.F.; Methodology, C.F. and V.V.; Project administration, G.B.; Supervision, M.G., A.D. and G.B.; Writing—Original draft, C.F. and V.V.; Writing—Review and editing, E.P., L.R. and A.B. All authors have read and agreed to the published version of the manuscript.

Funding: This research carried out in the framework of the nEU-Med project, *Origins of a new economic union (7th–12th centuries): resources, landscapes and political strategies in a Mediterranean region.* The project received fundings from the European Research Council (ERC) under the European Union's Horizon 2020 research and innovation program (grant agreement n. 670792).

Institutional Review Board Statement: Not applicable.

Informed Consent Statement: Not applicable.

Data Availability Statement: The data presented in this study are available on request from the corresponding author.

Acknowledgments: We are grateful to the entire nEU-Med research team for their kind assistance for all the scientific and logistic support during the fieldwork. Finally, our thanks go to the three anonymous referees for all their helpful suggestions and comments on the first draft of this paper.

Conflicts of Interest: The authors declare no conflict of interest.

References

1. Hein, A.; Kilikoglou, V. Compositional variability of archaeological ceramics in the eastern Mediterranean and implications for the design of provenance studies. *J. Archaeol. Sci. Rep.* **2017**, *16*, 564–572. [CrossRef]
2. Hayes, K. Parameters in the use of pXRF for archaeological site prospection: A case study at the Reaume Fort Site, Central Minnesota. *J. Archaeol. Sci.* **2013**, *40*, 3193–3211. [CrossRef]
3. Liritzis, I.; Xanthopoulou, V.; Palamara, E.; Papageorgiou, I.; Iliopoulos, I.; Zacharias, N.; Vafiadou, A.; Karydas, A.G. Characterization and provenance of ceramic artifacts and local clays from Late Mycenaean Kastrouli (Greece) by means of p-XRF screening and statistical analysis. *J. Cult. Herit.* **2020**, *46*, 61–81. [CrossRef]
4. Emmitt, J.J.; McAlister, A.J.; Phillipps, R.S.; Holdaway, S.J. Sourcing without sources: Measuring ceramic variability with pXRF. *J. Archaeol. Sci. Rep.* **2018**, *17*, 422–432. [CrossRef]
5. Cannavò, V.; Photos-Jones, E.; Levi, S.T.; Brunelli, D.; Fragnoli, P.; Lomarco, G.; Lugli, F.; Martinelli, M.C.; Sforna, M.C. p-XRF analysis of multi-period Impasto and Cooking Pot wares from the excavations at Stromboli-San Vincenzo, Aeolian Islands, Italy. *STAR Sci. Technol. Archaeol. Res.* **2017**, *3*, 326–333. [CrossRef]
6. LeMoine, J.B.; Halperin, C.T. Comparing INAA and pXRF analytical methods for ceramics: A case study with Classic Maya wares. *J. Archaeol. Sci. Rep.* **2021**, *36*, 102819. [CrossRef]
7. Bianchi, G.; Hodges, R. *Origins of a New Economic Union (7th-12th Centuries): Preliminary Results of the nEU-Med Project: October 2015-March 2017*; All'insegna del Giglio: Sesto Fiorentino, Italy, 2018.
8. Bianchi, G.; Hodges, R. *The nEU-Med Project: Vetricella, an Early Medieval Royal Property on Tuscany's Mediterranean*; All'insegna del Giglio: Sesto Fiorentino, Italy, 2020; ISBN 9788878149717.
9. Michałowski, A.; Niedzielski, P.; Kozak, L.; Teska, M.; Jakubowski, K.; Żółkiewski, M. Archaeometrical studies of prehistoric pottery using portable ED-XRF. *Meas. J. Int. Meas. Confed.* **2020**, *159*, 107758. [CrossRef]
10. Glascock, M.D.; Neff, H.; Stryker, K.S.; Johnson, T.N. Sourcing archaeological obsidian by an abbreviated NAA procedure. *J. Radioanal. Nucl. Chem.* **1994**, *180*, 29–35. [CrossRef]
11. Gratuze, B. Obsidian characterization by laser ablation ICP-MS and its application to prehistoric trade in the Mediterranean and the Near East: Sources and distribution of obsidian within the Aegean and Anatolia. *J. Archaeol. Sci.* **1999**, *26*, 869–881. [CrossRef]
12. Cochrane, E.E.; Neff, H. Investigating compositional diversity among Fijian ceramics with laser ablation-inductively coupled plasma-mass spectrometry (LA-ICP-MS): Implications for interaction studies on geologically similar islands. *J. Archaeol. Sci.* **2006**, *33*, 378–390. [CrossRef]
13. Palanivel, R.; Meyvel, S. Microstructural and microanalytical study-(SEM) of archaeological pottery artefacts. *Rom. J. Phys.* **2010**, *55*, 333–341.
14. Shackley, M.S. Gamma Rays, X-rays and Stone Tools: Some Recent Advances in Archaeological Geochemistry. *J. Archaeol. Sci.* **1998**, *25*, 259–270. [CrossRef]
15. Guerra, M.F. Analysis of Archaeological Metals. The Place of XRF and PIXE in the Determination of Technology and Provenance. *X-ray Spectrom. Int. J.* **1998**, *27*, 73–80. [CrossRef]
16. Hall, M.E. Pottery production during the Late Jomon period: Insights from the chemical analyses of Kasori B pottery. *J. Archaeol. Sci.* **2004**, *31*, 1439–1450. [CrossRef]
17. Charlton, M.F.; Crew, P.; Rehren, T.; Shennan, S.J. Explaining the evolution of ironmaking recipes—An example from northwest Wales. *J. Anthr. Archaeol.* **2010**, *29*, 352–367. [CrossRef]
18. Iñañez, J.G.; Buxeda, I.; Garrigós, J.; Speakman, R.J.; Glascock, M.D.; Suárez, E.S. Characterization of 15th-16th century majolica pottery found on the Canary Islands. Archaeological Chemistry: Analytical Methods and Archaeological Interpretation. In *Proceedings of the ACS Symposium Series*; American Chemical Society: Washington, DC, USA, 2007; Volume 968, pp. 376–398.
19. Speakman, R.J.; Little, N.C.; Creel, D.; Miller, M.R.; Iñañez, J.G. Sourcing ceramics with portable XRF spectrometers? A comparison with INAA using Mimbres pottery from the American Southwest. *J. Archaeol. Sci.* **2011**, *38*, 3483–3496. [CrossRef]
20. Forster, N.; Grave, P.; Vickery, N.; Kealhofer, L. Non-destructive analysis using PXRF: Methodology and application to archaeological ceramics. *X-ray Spectrom.* **2011**, *40*, 389–398. [CrossRef]
21. Goren, Y.; Mommsen, H.; Klinger, J. Non-destructive provenance study of cuneiform tablets using portable X-ray fluorescence (pXRF). *J. Archaeol. Sci.* **2011**, *38*, 684–696. [CrossRef]
22. Johnson, J. Accurate Measurements of Low Z Elements in Sediments and Archaeological Ceramics Using Portable X-ray Fluorescence (PXRF). *J. Archaeol. Method Theory* **2014**, *21*, 563–588. [CrossRef]
23. Hunt, A.M.W.; Speakman, R.J. Portable XRF analysis of archaeological sediments and ceramics. *J. Archaeol. Sci.* **2015**, *53*, 626–638. [CrossRef]
24. Holmqvist, E. Handheld Portable Energy-Dispersive X-Ray Fluorescence Spectrometry (pXRF). In *The Oxford Handbook of Archaeological Ceramic Analysis*; Oxford University Press: Oxford, UK, 2016; pp. 362–381.
25. Liritzis, I.; Zacharias, N. Portable XRF of archaeological artifacts: Current research, potentials and limitations. In *X-ray Fluorescence Spectrometry (XRF) in Geoarchaeology*; Springer: Berlin/Heidelberg, Germany, 2011; pp. 109–142, ISBN 9781441968852.
26. Tykot, R.H.; White, N.M.; Du Vernay, J.P.; Freeman, J.S.; Hays, C.T.; Koppe, M.; Hunt, C.N.; Weinstein, R.A.; Woodward, D.S. Advantages and disadvantages of pXRF for archaeological ceramic analysis: Prehistoric pottery distribution and trade in NW

Florida. In *Archaeological Chemistry VIII*; ACS Symposium Series; ACS Publications: Washington, DC, USA, 2013; Volume 1147, pp. 233–244.

27. Frahm, E.; Doonan, R.C.P. The technological versus methodological revolution of portable XRF in archaeology. *J. Archaeol. Sci.* **2013**, *40*, 1425–1434. [CrossRef]

28. Shugar, A.N.; Mass, J.L. *Handheld XRF for Art and Archaeology*; Leuven University Press: Leuven, Belgium, 2012; ISBN 9789461660695.

29. Shugar, A.N. Portable X-ray fluorescence and archaeology: Limitations of the instrument and suggested methods to achieve desired results. In Proceedings of the ACS Symposium Series. *Am. Chem. Soc.* **2013**, *1147*, 173–193.

30. Ceccarelli, L.; Rossetti, I.; Primavesi, L.; Stoddart, S. Non-destructive method for the identification of ceramic production by portable X-rays Fluorescence (pXRF). A case study of amphorae manufacture in central Italy. *J. Archaeol. Sci. Rep.* **2016**, *10*, 253–262. [CrossRef]

31. Aimers, J.J.; Farthing, D.J.; Shugar, A.N. Handheld XRF analysis of Maya ceramics: A pilot study presenting issues related to quantification and calibration. In *Handheld XRF for Art and Archaeology*; Leuven University Press: Leuven, Belgium, 2012; pp. 423–448, ISBN 9789461660695.

32. Speakman, R.J.; Shackley, M.S. Silo science and portable XRF in archaeology: A response to Frahm. *J. Archaeol. Sci.* **2013**, *40*, 1435–1443. [CrossRef]

33. Conrey, R.M.; Goodman-Elgar, M.; Bettencourt, N.; Seyfarth, A.; Van Hoose, A.; Wolff, J.A. Calibration of a portable X-ray fluorescence spectrometer in the analysis of archaeological samples using influence coefficients. *Geochem. Explor. Environ. Anal.* **2014**, *14*, 291–301. [CrossRef]

34. Ceccarelli, L. Production and Trade in Central Italy in the Roman Period: The Amphora Workshop of Montelabate in Umbria. *Pap. Br. Sch. Rome* **2017**, *85*, 109–141. [CrossRef]

35. Craig, N.; Speakman, R.J.; Popelka-Filcoff, R.S.; Glascock, M.D.; Robertson, J.D.; Shackley, M.S.; Aldenderfer, M.S. Comparison of XRF and PXRF for analysis of archaeological obsidian from southern Perú. *J. Archaeol. Sci.* **2007**, *34*, 2012–2024. [CrossRef]

36. Shackley, M.S. An introduction to X-ray fluorescence (XRF) analysis in archaeology. In *X-ray Fluorescence Spectrometry (XRF) in Geoarchaeology*; Springer: Berlin/Heidelberg, Germany, 2011; pp. 7–44, ISBN 9781441968852.

37. Goodale, N.; Bailey, D.G.; Jones, G.T.; Prescott, C.; Scholz, E.; Stagliano, N.; Lewis, C. PXRF: A study of inter-instrument performance. *J. Archaeol. Sci.* **2012**, *39*, 875–883. [CrossRef]

38. Frahm, E. Ceramic studies using portable XRF: From experimental tempered ceramics to imports and imitations at Tell Mozan, Syria. *J. Archaeol. Sci.* **2018**, *90*, 12–38. [CrossRef]

39. Shackley, M.S. Portable X-ray fluorescence spectrometry (pXRF): The good, the bad, and the ugly. *Archaeol. Southwest Mag.* **2012**, *26*, 1–8.

40. Papachristodoulou, C.; Gravani, K.; Oikonomou, A.; Ioannides, K. On the provenance and manufacture of red-slipped fine ware from ancient Cassope (NW Greece): Evidence by X-ray analytical methods. *J. Archaeol. Sci.* **2010**, *37*, 2146–2154. [CrossRef]

41. Gajić-Kvaščev, M.D.; Marić-Stojanović, M.D.; Jančić-Heinemann, R.M.; Kvaščev, G.S.; Andrić, V.D. Non-destructive characterisation and classification of ceramic artefacts using pEDXRF and statistical pattern recognition. *Chem. Cent. J.* **2012**, *6*, 102. [CrossRef] [PubMed]

42. López-García, P.; Argote-Espino, D.; Fačevicová, K. Statistical processing of compositional data. The case of ceramic samples from the archaeological site of Xalasco, Tlaxcala, Mexico. *J. Archaeol. Sci. Rep.* **2018**, *19*, 100–114. [CrossRef]

43. Baxter, M.J.; Buck, C.E. Data handling and statistical analysis. In *Modern Analytical Methods in Art and Archaeology*; Wiley: Hoboken, NJ, USA, 2000; pp. 681–746.

44. Aitchison, J.; Barceló-Vidal, C.; Pawlowsky-Glahn, V. Some comments on compositional data analysis in archaeometry, in particular the fallacies in Tangri and Wright's dismissal of logratio analysis. *Archaeometry* **2002**, *44*, 295–304. [CrossRef]

45. Filzmoser, P.; Hron, K.; Reimann, C. Principal component analysis for compositional data with outliers. *Off. J. Int. Environ. Soc.* **2009**, *20*, 621–632. [CrossRef]

46. Bergman, J.; Lindahl, A. Optimising archaeologic ceramics h-XRF analyses. In *International Workshop on Compositional Data Analysis*; Springer: Cham, Switzerland, 2016.

47. Mauran, G.; Caron, B.; Détroit, F.; Nankela, A.; Bahain, J.-J.; Pleurdeau, D.; Lebon, M. Data pretreatment and multivariate analyses for ochre sourcing: Application to Leopard Cave (Erongo, Namibia). *J. Archaeol. Sci. Rep.* **2021**, *35*, 102757. [CrossRef]

48. Baxter, M.J.; Freestone, I.C. Log-ratio compositional data analysis in Archaeometry. *Archaeometry* **2006**, *48*, 511–531. [CrossRef]

49. Tanasi, D.; Tykot, R.H.; Pirone, F.; McKendry, E. Provenance study of prehistoric ceramics from sicily: A comparative study between pXRF and XRF. *Open Archaeol.* **2017**, *3*, 222–234. [CrossRef]

50. Iserlis, M.; Steiniger, D.; Greenberg, R. Contact between first dynasty Egypt and specific sites in the Levant: New evidence from ceramic analysis. *J. Archaeol. Sci. Rep.* **2019**, *24*, 1023–1040. [CrossRef]

51. Tite, M.S. Ceramic production, provenance and use A review. *Archaeometry* **2008**, *50*, 216–231. [CrossRef]

52. Speakman, R.J.; Neff, H. *Laser Ablation ICP-MS in Archaeological Research*; UNM Press: Albuquerque, NM, USA, 2005.

53. Odelli, E.; Selvaraj, T.; Perumal, J.; Palleschi, V.; Legnaioli, S.; Raneri, S. Pottery production and trades in Tamil Nadu region: New insights from Alagankulam and Keeladi excavation sites. *Herit. Sci.* **2020**, *8*, 56. [CrossRef]

54. Borges, C.S.; Weindorf, D.C.; Nascimento, D.C.; Curi, N.; Guilherme, L.R.G.; Carvalho, G.S.; Ribeiro, B.T. Comparison of portable X-ray fluorescence spectrometry and laboratory-based methods to assess the soil elemental composition: Applications for wetland soils. *Environ. Technol. Innov.* **2020**, *19*, 100826. [CrossRef]

55. Montana, G. Ceramic raw materials: How to recognize them and locate the supply basins—Mineralogy, petrography. *Archaeol. Anthropol. Sci.* **2020**, *12*, 175. [CrossRef]

56. Cantini, F.; Grassi, F. Produzione, circolazione e consumo della ceramica in Toscana tra la fine del X e il XIII secolo. In Proceedings of the S. Gelichi (a cura di), Atti del IX Congresso Internazionale sulla ceramica medievale nel Mediterraneo, Venezia, Italy, 22–27 October 2012; pp. 131–140.

57. Basile, L.; Grassi, F.; Riccardi, M.P. Basso Gli scarichi di fornace di Roccastrada (Gr): Nuove analisi archeologiche ed archeometriche. In *LRCW3 Late Roman Coarse Wares*; Menchelli, S., Santoro, S., Pasquinucci, M., Guiducci, G., Eds.; BAR: Oxford, UK, 2011; p. S2185.

58. Grassi, F. Production, consumption and political complexity: Early Medieval pottery in Castile and southern Tuscany (7th–10th centuries). In *Social Complexity in Early Medieval Rural Communities. The North-Western Iberia Archaeological Record*; Archaeopress: Oxford, UK, 2016; pp. 91–112.

59. Fornacelli, C.; Briano, A.; Chiarantini, L.; Bianchi, G.; Benvenuti, M.; Giamello, M.; Kang, J.S.; Villa, I.M.; Talarico, F.M.; Hodges, R. Archaeometric Provenance Constraints for Early Medieval Sparse Glazed Pottery from Donoratico (Livorno, Italy). *Archaeometry* **2021**, *63*, 549–576. [CrossRef]

60. Ponta, E. *Il Paesaggio e le sue Trasformazioni tra IV e VIII sec d.C. tra Costa ed Entroterra. Il Caso Della Toscana Centro Meridionale*; University of Pisa: di Pisa, Italy, 2019.

61. Russo, L. *Ceramica Grezza, Depurata e Semidepurata: Produzione, Funzione e Circolazione in un Territorio Della Toscana Sud-Occidentale*; Colline Metallifere e territori limitrofi tra VIII- e XI secolo; Università degli Studi di Siena: Siena, Italy, 2021.

62. Grassi, F. *Gli Apparati Produttivi, i Mercati ed il Consumo Della Ceramica Nella TOscana Meridionale (VIII-XIV secolo): Il Confronto Tra i Siti Rurali ed Urbani*; Università degli Studi di Siena: Siena, Italy, 2005.

63. Vaccaro, E. *Sites and Pots: Settlement and Economic Patterns in Southern Tuscany (AD 300-900)*; Archaeopress: Oxford, UK, 2012.

64. Bianchi, G. Analyzing fragmentation in the Early Middle Ages: The Tuscan model and the countryside in central-northern Italy. In *New Directions in Early Medieval European Archaeology: Spain and Italy Compared*; Brepols: Turnhout, Belgium, 2015; pp. 301–334.

65. Marasco, L. Un castello di pianura in località Vetricella a Scarlino (Scarlino Scalo, GR): Indagini preliminari e saggi di verifica. In *P. FAVIA, G. VOLPE (a cura di), V Congresso Nazionale di Archeologia Medievale*; Favia, P., Volpe, G., Eds.; University of Siena: Siena, Italy, 2009; pp. 326–331.

66. Carmignani, L.; Conti, P.; Cornamusini, G.; Meccheri, L. The internal Northern Apennines, the northern Tyrrhenian Sea and the Sardinia-Corsica block. *Geol. Italy Spec. Vol. Ital. Geol. Soc. IGC* **2004**, *32*, 59–77.

67. Gianelli, G.; Manzella, A.; Puxeddu, M. Crustal models of the geothermal areas of southern Tuscany (Italy). *Tectonophysics* **1997**, *281*, 221–239. [CrossRef]

68. Brunet, C.; Monié, P.; Jolivet, L.; Cadet, J.-P. Migration of compression and extension in the Tyrrhenian Sea, insights from 40Ar/39Ar ages on micas along a transect from Corsica to Tuscany. *Tectonophysics* **2000**, *321*, 127–155. [CrossRef]

69. Dini, A.; Gianelli, G.; Puxeddu, M.; Ruggieri, G. Origin and evolution of Pliocene-Pleistocene granites from the Larderello geothermal field (Tuscan Magmatic Province, Italy). *Lithos* **2005**, *81*, 1–31. [CrossRef]

70. Viti, C.; Lupieri, M.; Reginelli, M. Weathering sequence of rhyolitic minerals: The kaolin deposit of Torniella (Italy). *Neues Jahrb. Für Miner.-Abh.* **2007**, *183*, 203–213. [CrossRef]

71. Lattanzi, P.; Benvenuti, M.; Costagliola, P.; Tanelli, G. An overview on recent research on the metallogeny of Tuscany, with special reference to Apuane Alps. *Mem. Della Soc. Geol. Ital.* **1994**, *48*, 613–625.

72. Bonciani, F.; Callegari, I.; Conti, P.; Cornamusini, G.; Carmignani, L. Neogene post-collisional evolution of the internal Northern Apennines: Insights from the upper Fiora and Albegna valleys (Mt. Amiata geothermal area, southern Tuscany). *Boll. Della Soc. Geol. Ital.* **2005**, *3*, 103–118.

73. Bossio, A.; Costantini, A.; Foresi, L.M.; Lazzarotto, A.; Mazzanti, R.; Mazzei, R.; Pascucci, V.; Salvatorini, G.; Sandrelli, F.; Terzuoli, A.; et al. Neogene-Quaternary sedimentary evolution in the western side of the Northern Apennines (Italy). *Mem. Della Soc. Geol. Ital.* **1998**, *52*, 513–525.

74. Lattanzi, P.; Benvenuti, M.; Costagliola, P.; Maineri, C.; Mascaro, I.; Tanelli, G.; Dini, A.; Ruggieri, G. Magmatic versus hydrothermal processes in the formation of raw ceramic material deposits in Southern Tuscany. In Proceedings of the Tenth International Symposium On Water-Rock Interaction, Villasimius, Italy, 10–15 July 2001; pp. 725–728.

75. Stoltman, J.B. The Role of Petrography in the Study of Archaeological Ceramics. In *Earth Sciences and Archaeology*; Springer: Boston, MA, USA, 2001; pp. 297–326.

76. Courty, M.A.; Goldberg, P.; Macphail, R.I. Soil micromorphology in archaeology. *Endeavour* **1990**, *14*, 163–171. [CrossRef]

77. Killick, D. The awkward adolescence of archaeological science. *J. Archaeol. Sci.* **2015**, *56*, 242–247. [CrossRef]

78. Hein, A. Revisiting the groups: Exploring the feasibility of portable EDXRF in provenance studies of transport amphorae in the Eastern Aegean. In *Application of Portable Energy-Dispersive X-ray Fluorescence to the Analysis of Archaeological Ceramics and Glass. Topoi. Berlin Studies of the Ancient World*; Peking University Press: Beijing, China, 2017.

79. Hein, A.; Dobosz, A.; Day, P.M.; Kilikoglou, V. Portable ED-XRF as a tool for optimizing sampling strategy: The case study of a Hellenistic amphora assemblage from Paphos (Cyprus). *J. Archaeol. Sci.* **2021**, *133*, 105436. [CrossRef]

80. Kenna, T.C.; Nitsche, F.O.; Herron, M.M.; Mailloux, B.J.; Peteet, D.; Sritrairat, S.; Sands, E.; Baumgarten, J. Evaluation and calibration of a Field Portable X-Ray Fluorescence spectrometer for quantitative analysis of siliciclastic soils and sediments. *J. Anal. At. Spectrom.* **2011**, *26*, 395–405. [CrossRef]

81. Walton, M.S. *Materials Chemistry Investigation of Archaeological Lead Glazes*; University of Oxford: Oxford, UK, 2004.

82. Téreygeol, F.; Henning, J. Production and circulation of silver and secondary products (lead and glass) from Frankish royal silver mines at Melle (eighth to tenth century). In *Volume 1 The Heirs of the Roman West*; Walter de Gruyter: Berlin, Germany, 2007; Volume 5, p. 123.

83. Niederschlag, E.; Pernicka, E.; Seifert, T.; Bartelheim, M. The Determination of Lead Isotope Ratios by Multiple Collector Icp-Ms: A Case Study of Early Bronze Age Artefacts and their Possible Relation With Ore Deposits of the Erzgebirge. *Archaeometry* **2003**, *45*, 61–100. [CrossRef]

84. Asmus, B. *Medieval Copper Smelting in the Harz Mountains, Germany*; University College London: London, UK, 2011.

85. Costa, V.; Urban, F. Lead and its alloys: Metallurgy, deterioration and conservation. *Stud. Conserv.* **2005**, *50*, 48–62. [CrossRef]

86. Salminen, R. *Geochemical Atlas of Europe*; Geological Survey of Finland: Espoo, Finland, 2005.

87. Glascock, M. Characterization of Archaeological Ceramics at MURR by Neutron Activation Analysis and Multivariate Statistics. In *Chemical Characterization of Ceramic Pastes in Archaeology*; Neff, H., Ed.; Prehistory Press: Madison, WI, USA, 1992; pp. 11–26.

88. Baxter, M.J.; Cool, H.E.M.; Heyworth, M.P.; Jackson, C. Compositional Variability in Colourless Roman Vessel Glass. *Archaeometry* **1995**, *37*, 129–141. [CrossRef]

89. Aitchison, J. Lograntios and natural laws in compositional data analysis. *Math. Geol.* **1999**, *31*, 563–580. [CrossRef]

90. Beardah, C.C.; Baxter, M.J.; Cool, H.E.M.; Jackson, C.M. Compositional Data Analysis of Archaeological Glass: Problems and Possible solutions. In Proceedings of the Compositional Data Analysis Workshop, CODAWORK, Girona, Spain, 15–17 October 2003; pp. 279–300.

91. Simsek Franci, G. Handheld X-ray Fluorescence (XRF) Versus Wavelength Dispersive XRF: Characterization of Chinese Blue-and-White Porcelain Sherds Using Handheld and Laboratory-Type XRF Instruments. *Appl. Spectrosc.* **2020**, *74*, 314–322. [CrossRef]

92. Demirsar Arli, B.; Simsek Franci, G.; Kaya, S.; Arli, H.; Colomban, P. Portable X-ray Fluorescence (p-XRF) Uncertainty Estimation for Glazed Ceramic Analysis: Case of Iznik Tiles. *Heritage* **2020**, *3*, 1302–1329. [CrossRef]

93. Kuleff, I.; Djingova, R. Provenance study of pottery; choice of elements to be determined. *ArchéoSci. Rev. D'archéom.* **1996**, *20*, 57–67. [CrossRef]

94. Annis, M.B. Ceramica altomedievale a vetrina pesante e ceramica medievale a vetrina sparsa proveniente dallo scavo di San Sisto Vecchio in Roma: Analisi tecnologica e proposta interpretativa. In Proceedings of the Ceramica Invetriata Tardoantica e Altomedievale in Italia, Atti del Seminario, Siena, Italy, 23–24 February 1990; pp. 394–417.

95. Sfrecola, S. La ceramica a vetrina pesante (Forum Ware) e la ceramica a vetrina sparsa da alcuni siti nella Campagna Romana. In Proceedings of the Ceramica Invetriata Tardoantica e Altomedievale in Italia, Atti del Seminario, Siena, Italy, 23–24 February 1990; pp. 562–579.

96. Tassi, E.; Grifoni, M.; Bardelli, F.; Aquilanti, G.; La Felice, S.; Iadecola, A.; Lattanzi, P.; Petruzzelli, G. Evidence for the natural origins of anomalously high chromium levels in soils of the Cecina Valley (Italy). *Environ. Sci. Process. Impacts* **2018**, *20*, 965–976. [CrossRef] [PubMed]

97. Vaccaro, E. Long-distance ceramic connections: Portus Scabris (Portiglioni, GR), coastal tuscany and the Tyrrhenian Sea. In *Origins of a New Economic Union (7th–12th Centuries). Preliminary Results of the nEU-Med Project (October 2015–March 2017)*; All'Insegna del Giglio: Firenze, Italy, 2018; pp. 81–99.

Article

Archaeological and Chemical Investigation on the High Imperial Mosaic Floor Mortars of the Domus Integrated in the Museum of Archaeology D. Diogo de Sousa, Braga, Portugal

Ana Fragata [1,*], Jorge Ribeiro [2], Carla Candeias [1], Ana Velosa [3] and Fernando Rocha [1]

[1] GeoBioTec, Geosciences Department, Campus de Santiago, University of Aveiro, 3810-193 Aveiro, Portugal; candeias@ua.pt (C.C.); tavares.rocha@ua.pt (F.R.)

[2] Lab2PT, Institute of Social Sciences, Campus de Gualtar, University of Minho, 4710-057 Braga, Portugal; jribeiro@uaum.uminho.pt

[3] RISCO, Department of Civil Engineering, Campus de Santiago, University of Aveiro, 3810-193 Aveiro, Portugal; avelosa@ua.pt

* Correspondence: afragata@ua.pt

Abstract: This paper intends to characterize the floor mortar layers (*nucleus*, *rudus* and *statumen*) of the high imperial mosaics of the *domus* integrated in the Museum of Archeology D. Diogo de Sousa, the oldest roman housing testimonies known in Braga, Portugal. It offers an important archaeological and historical contextualization and first chemical characterization attempt on the mortars. The study of 13 mortar samples was carried out at a chemical level through X-ray fluorescence spectroscopy (XRF). All samples presented low lime content when compared to similar studies. A high chemical similarity between nucleus mortars (*opus signinum*) and chemical composition differences between *rudus* and *statumen* mortars was determined, confirmed by statistical analyses. Their composition was distinctly related to the stratigraphic position of each floor mortar layer, following Vitruvius' model, and to the external conditions and treatments (e.g., capillary rise with soluble salts and application of chemical treatments), to which they were submitted.

Keywords: roman mortar; historic mortar; mosaic floors; XRF; *Bracara Augusta*

Citation: Fragata, A.; Ribeiro, J.; Candeias, C.; Velosa, A.; Rocha, F. Archaeological and Chemical Investigation on the High Imperial Mosaic Floor Mortars of the Domus Integrated in the Museum of Archaeology D. Diogo de Sousa, Braga, Portugal. *Appl. Sci.* **2021**, *11*, 8267. https://doi.org/10.3390/app11178267

Academic Editors: Anna Galli and Letizia Bonizzoni

Received: 5 August 2021
Accepted: 31 August 2021
Published: 6 September 2021

Publisher's Note: MDPI stays neutral with regard to jurisdictional claims in published maps and institutional affiliations.

1. Introduction

Roman mortars are noted for their high durability and complex technological knowledge, related with its composition and execution methods. Their study provides fundamental information regarding the knowledge of the archaeological sites. Some studies have been developed on the characterization of roman mortars in Portuguese archaeological sites, such as Beja-Pisões, Tróia, Conimbriga and Marvão-Anmaia [1–4]. The roman floor mosaics' substrate has several preparatory mortar layers, carefully built (in terms of number, thickness and composition). Specifically, these kind of mortar substrates are under study elsewhere, such as Greece, Italy, Slovenia and Spain [5–12]. However, in *Bracara Augusta*, the Roman name of Braga (Portugal), the roman mortars from its archaeological sites are poorly analyzed and there is no known bibliographic references concerning their characterization.

The Roman city of *Bracara Augusta*, and its evolution between the end of the first century BC and Late Antiquity (V-VII centuries) is known based on the numerous remains recovered by urban archaeology over the last 40 years, in the context of the "Campo Arqueológico de Braga" (CAB; Braga archeological Campus) and the *Bracara Augusta* Project, created in 1976 [13], within the scope of preventive archaeology interventions, conducted by the Archaeology Unit of the University of Minho (UAUM) and the Archaeology Office of the Municipality of Braga (GACMB). In this process, the role of the Museum of Archaeology D. Diogo de Sousa (MDDS), by its contribution in supporting research to the defense and preservation of the archaeological heritage of Braga, must be highlighted. Braga is a city with more than 2000 years of history, founded by the emperor August, around 16–15 BC, in the context of the administrative reorganization of the Northern Hispania [14,15]. *Bracara Augusta* was one of the most important Roman cities of the *Hispania* northwest, capital of *Conventus*, in the High Empire, of the *Gallaecia*, under Diocletian, and of the Suebic Kingdom, roughly from 411 AD.

Since 1976, dozens of archaeological excavations allowed us to understand the evolution of the city history and the knowledge about its construction techniques and employed materials. In the Stables (Cavalariças, in Portuguese) Archaeological Site of Braga (CVL), in 1991, two mosaic floors were exhumed, covering two compartments belonging to one of the *domus* located to the south of the city *forum*.

1.1. Archaeological Background

Bracara Augusta was an ex nuovo foundation, perfectly planned, and its urban fabric has been recognized through excavations carried out since the 70's [14,15]. The Stables Archaeological Site of the Braga, also known as Zone P1 [16], is located in the southern half of the city, in a prime area (south of the *forum*) on both sides of the maximum *kardo*, east of the blocks occupied by the group of best-known public buildings in the city: the Roman Baths of Alto da Cividade and the Theatre (Figure 1) [17,18].

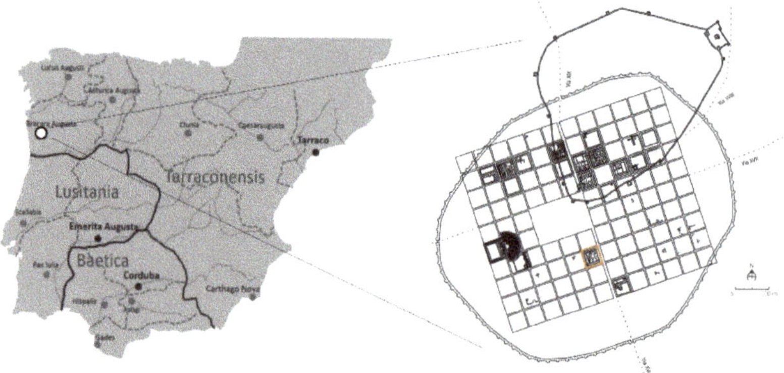

Figure 1. Location of the Stables Archaeological Site in the plan of the roman city, low imperial and medieval walls. Adapted from [19].

This is the largest excavated area in the southern part of the ancient Roman city and currently houses the facilities of the MDDS, open to the public since 2006. In this place, there were two rectangular buildings, used as stables, dating from the beginning of the 20th century, and located at an elevation of approximately 184 m above sea level. These buildings are surrounded in the west by the Bombeiros Voluntários Street and in the east by an unbuilt area [19,20]. The archaeological work was spread over four excavation campaigns that took place between 1986 and 2002. The first intervention occurred in 1986, to assess the archaeological potential of the area (three surveys) [16]. Between 1988 and 1989, there was a second campaign, covering the entire area of the Museum implantation (24 surveys) [20], followed by a third one, in 1996, extending the intervention to the Museum's gardens [19]. Finally, the fourth and last campaign took place in 2002, at a time when the Museum was already built, in the gardens, on the eastern part of the building [19]. The work carried out allowed the identification of the archaeological remains belonging to several *insulae* of the Roman city, some of which had housing functionalities, chronologically set between the 1st and the 7th century, allowing us to understand the evolution of the city in the blocks located to the south of it. Therefore, it is a large area, which was left out of the medieval city, having certainly been transformed into agricultural spaces, which contributed to the preservation of the existing ruins. Testimonies of the various stages of occupation were found, such as pavements of *opus signinum*, a pavement of carved stones in the shape of diamonds and rectangles, various mosaics, and walls [16,19,20]. The interpretation of this set allowed the identification of two blocks of the Roman city, designated, respectively, by A and B arranged on each side of the south segment of the maximum *kardo*. Block A was occupied by a *domus*, from which most of the central structures were recovered. From block B, the data collected was more dispersed and complex, and did not allow for a reliable interpretation [19]. The essential part of the exhumed archaeological remains is now buried, apart from the mosaic targeted by this study, which formalized the soil of a room in a house, preserved and musealized in the basement of the MDDS service block [16,20].

1.2. The Bracara Augusta Mosaics and the Findings of the Stables Archaeological Site

Braga's mosaics were studied by Abraços [21,22], in the coordination of a large team. The first set identified was discovered in the "Campo das Carvalheiras" area, in the context of the construction of the new Orphans Seminary; however, most of the documented elements result from the work carried out since 1976, with the formation of the CAB and the UAUM. The large number of archaeological excavations carried out by the UAUM have allowed the identification, registration, safeguarding and collection of materials and a better understanding of the *Bracara Augusta* mosaics. The city has been known, throughout its history, by diverse periods of building construction and economic growth [14,15,23], associated with successive changes in its status, which manifested themselves in various programs for the city's beautification, which integrates the mosaic decoration of many of its public and private buildings. Few of the city's mosaics remain in situ, the majority being deposited in MDDS [22]. A set of 69 mosaic coverings, including elements from Dume and Falperra, essentially decorated with geometric motifs, and some sets decorated with aquatic motifs and vases, were documented. Mosaics from the 1st century are known, however most of these panels date from the 3rd to 4th centuries, as a result of the intense construction and urban renewal program that the city experienced by that time [22].

With specific regard to the mosaics of the CVL archaeological site, there are five records. The first identified set was in 1991, composed of two panels dating from the 1st century, which covered the floor of two rooms of a housing complex. The second, later, dating from the Suevo-Visigothic period, was identified in 1986 in the Museum gardens, and matches a pavement in *opus sectile* and *opus tesselatum*, both preserved in situ. Three fragments were also discovered in works carried out in 1988, testimonies of the *opus tesselatum* technique, whose dating was not possible to specify, given the size of the archaeological discovery and context [22].

The mosaic pavement, analyzed in this present study, and the associated structures, such as walls and pipes, were in a reasonable state of conservation [24]. The rarity of this type of mosaic in the region is due to its size and constitution, and the soil acidity that does not favors its conservation, besides the fact that is considered the oldest Roman housing testimonies known until now in Braga [25], dating from the August period.

1.3. The Presence of Marble and Lime in Bracara Augusta

The raw material most used by the builders of the Roman city was granite, an abundant rock-type in this region (Figure 2).

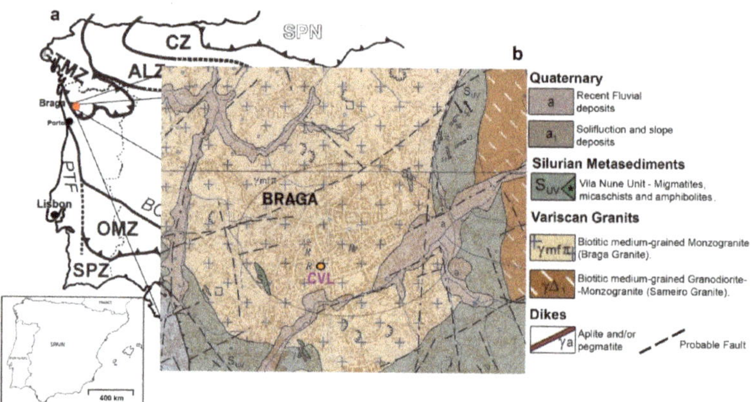

Figure 2. (**a**) Geological setting of the study area in the Iberian Peninsula; (**b**) Braga and the Stables Archaeological site geological context [26]. Adapted from [27].

Furthermore, several types of Braga's ceramics [28], as well as lime and marble [29] were imported from other places in the Roman Empire, such as Lusitania. The importance of marble, one of the most relevant stone resources in the Roman period, from the Marble Triangle (Estremoz, Borba, Vila Viçosa—Portugal), to *Asturica Augusta*, was studied in [30], and referenced therein. These being the marble suppliers of *Bracara Augusta*. However, as documented by a study of the city's architectural elements [31], in a universe of 356 analyzed pieces, only six were in marble. In addition to the above, its absence may also be justified by the costs associated with its acquisition or, possibly, an economy of reuse, very common in the Roman times and in the medieval world, in the context of the construction of new buildings, or even its reduction to obtain lime, an element of great importance in the Middle Ages. Similar to what was documented in the excavations of the Saint Raymond Museum, in Toulouse [32], there would be kilns for reducing marble pieces, including statuary, to obtain lime. Additionally, the assumptions related to the cost of the material and its reuse are in fact admissible, however the importation of marble and lime to Braga cannot be excluded and could have been superior to the scarce archaeological evidence available.

2. Description and Conservation of the Mosaic's Pavement

The construction techniques and materials used by the Romans were described, in the 1st century AC, by Vitruvius, in the work entitled "De Architectura" [33], according to the following down to top layers (Figure 3): (a) a first preparatory layer (*statumen*), consisting of rolled pebbles and stones, measuring about 12 cm (the size of a fist); (b) over, the *rudus* layer was placed, consisting of 3/4 of sand or gravel and small pebbles and $\frac{1}{4}$ of lime in a layer with ~22 cm (3/4 of a roman foot), to protect the mosaic against humidity and infiltrations; (c) the *nucleus*, a *opus signinum* layer, consisting of a thin lime mortar with fragments of tile, or ground brick, to level the floor to receive the *tesserae*, in a 1.5 cm layer; finally, (d) the *tesserae* was placed over the *nucleus*, in a layer of 2 to 3 mm, consisting of lime and very fine marble dust, that filled the interstices.

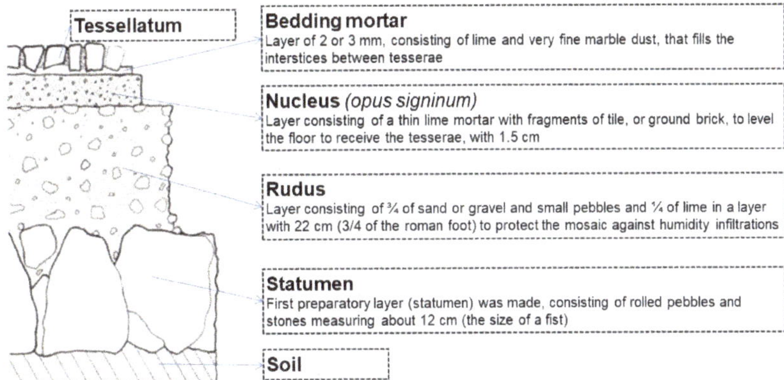

Figure 3. Stratigraphy of Roman mosaic floor according to Vitruvius' description [33].

Many of these Vitruvius model mosaics have subsisted until our days, but this structure was not always strictly respected, leading to a faster deterioration of the mosaic pavement.

In Figure 4, a section of the pavement of the mosaic of the MDDS crypt was drawn, in which can be observed the local adaptation of the different stages of the mosaic construction according to the Vitruvian model. In this case, not all the layers that support the mosaic respect the thickness of the Vitruvius' model. In Figure 5 it is shown the layers corresponding to the *nucleus, rudus* and *statumen* of the MDDS crypt mosaic floor.

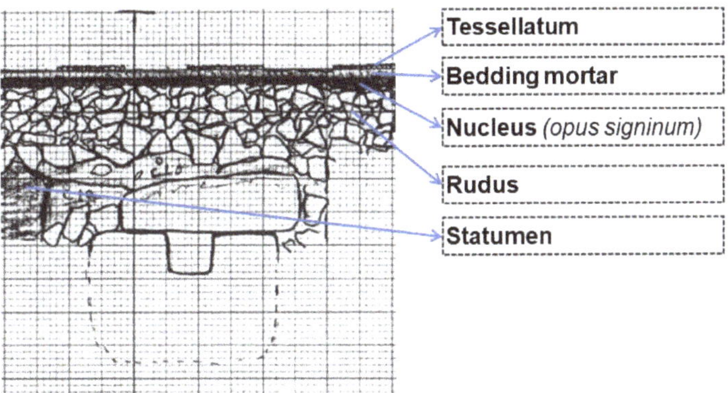

Figure 4. Section of the East profile of the mosaic floor of the Museum crypt [24,34].

Figure 5. Mosaic floor layers—*nucleus* and *rudus* (**left**) and statumen (**right**). (Photos generously provided by MDDS).

The protection of the area of the mosaic structure was carried out during the building construction for prior conservation actions (Figure 6a), although it was not efficient in controlling successive floods due to the low quota, at groundwater level, in comparison to the top of the hill located to the west—Hill of Cividade. Archeologists decided, to improve the protection of the area, to bury the mosaic floors until 1997, when the construction of the space that houses the mosaics was finished. The protective materials were removed (except the thin layer of sand that covered the mosaic floors) and some uncovered structures were consolidated and restored with a mortar based on hydraulic lime and gravel and a biocide, applied in the areas with biological colonization. A year later, when the sand was removed it was observed that the *tesserae* as well as other structures (walls, pipes and mortar pavements) were in a reasonable state of conservation. In 2003 and 2004, in order to drain the excess of water that reached the mosaic floors, a gallery was built, although it was not efficient. The floor mosaics were submitted to several treatments to remove the biological colonization [24,33], however it is still present (Figure 6b) due to room unfavorable humidity and temperature conditions. The anomalies in the structure of the crypt's mosaics may be the result of the high accumulation of water due to: (i) capillary rise of groundwater, likely through the more permeable materials of the mosaic structure; and (ii) rising damp due to the accumulation of surface water near the structure surrounding the mosaics. The lack of drainage system around the mosaic floor and lack of ventilation had unfavorable results on the mosaics [35,36].

(a)　　　　　　　　　　(b)

Figure 6. (**a**) Protection of the area of the structure of the mosaic during the building construction (photo gently provided by MDDS); (**b**) mosaic floor integrated in the crypt of the Museum of Archeology D. Diogo de Sousa.

The mosaic 1 feature squares were made of granite *tesserae* alternated with squares made of white limestone *tesserae* (Figure 7). The tray included, to this end, a drainage opening also covered with mosaic but with small squares with about 8 to 9 cm on the side. Mosaic 2 was decorated with squares of hourglass lines, equally made with granite and limestone *tesserae* (Figure 7). The mosaic covered a room whose maximum dimensions are 2.23 m in width and 3.22 m in length, with squares of approximately 20 cm on each side. In both cases, the limestone *tesserae* is poorly preserved [22] and only the granitic features remain preserved. The characterization of the *tesserae* composition determined that the white ones are composed of limestone, and the dark ones by granite, possibly pink granite from the Conde area, in Braga, assuming that this mosaic was made with local raw material [24].

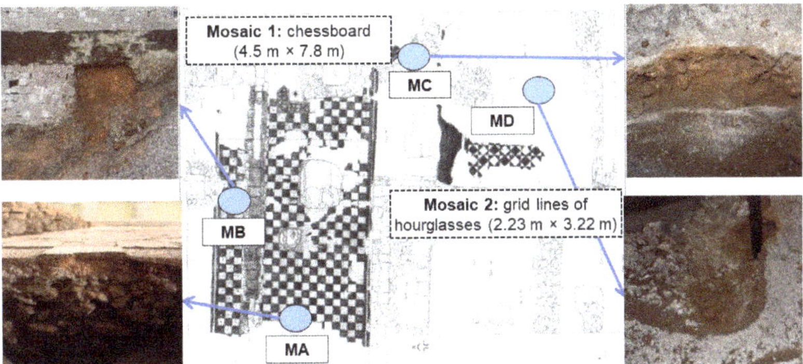

Figure 7. Schematic representation of the mosaic floor of the Museum crypt and sampling areas (Schematic representation gently provided by MDDS).

The durability of the floor mortars that supports those mosaic pavements is remarkable, although some conservation actions may be needed, as some cohesion loss was observed in those renders. No prior studies considering chemical characterization of those mortars in Braga were found. To achieve this, a study on those mortars was handle in this paper.

3. Materials and Methods

3.1. Materials

The study was conducted in the four different areas of the mosaic floor integrated in the crypt of the MDDS building. Micro-sampling was performed and carefully removed to prevent contamination and following the protocol of the technicians of MDDS.

This investigation was based on the characterization of 13 samples collected from the different mortar layers of the mosaic floor considering four different areas (MA, MB, MC, MD) of the mosaic floors (Figure 7). The samples were collected from original and well-preserved floor mortars (although some cohesion loss was observed during sampling, mainly in nucleus layer) and were macroscopically grouped, considering the color and the removal depth, in five different layers: layer 1—five samples from *nucleus* upper zone (MA2, MA3, MB1, MC1, MD1), layer 2—five samples from *nucleus* lower zone (MB2, MC2, MC3, MC4, MD2), layer 3—one sample between *nucleus* and *rudus* (MA4), layer 4—one sample from *rudus* (MD3) and layer 5—one sample from *statumen* (MA1) (Table 1).

The bedding mortar, a very thin layer consisting of lime and very fine marble (Figure 2), was not visible and, if it existed, may be below the original *tesselatum*; it was not possible to collect samples from this layer. Samples from the *nucleus* layer were removed from the *nucleus* upper layer (layer 1, close to the outer surface that previously was in contact with bedding mortar) and from the *nucleus* lower layer (layer 2, close to the *rudus* layer) to investigate compositional differences. Additionally, it was only possible to collect one sample from the interface between *nucleus* and *rudus* (MA4), and from *rudus* (MD3) *and* from *statumen* (MA1) layers; these samples need to be removed from deep and only one zone at the mosaics' floor mortars was defined by the museum technicians, to avoid as much damage as possible. The thickness of floor mortar layers varies. During nucleus sampling it was difficult to identify which group the samples belonged to in some cases.

Table 1. Studied samples and their characteristics.

Layer 1: Nucleus—Opus Signinum (Upper Zone of the Layer)				
Description	Removal depth: ~ 1.0/3.0 cm; color: reddish.			
Sample	MA2	MA3	MB1	MC1	MD1

Layer 2: Nucleus—Opus Signinum (Lower Zone of the Layer)				
Description	Removal depth: ~3.0/5.0 cm; color: reddish.			
Sample	MB2	MC2	MC3	MC4	MD2

Layer	Layer 3: Between Nucleus and Rudus	Layer 4: Rudus	Layer 5: Statumen
Description	Removal depth: ~5 cm; color: pinkish	Removal depth: ~5/10 cm; color: brown	Removal depth: ~25/30 cm; color: dark brown
Sample	MA4	MD3	MA1

3.2. Methods

This present study aims to characterize the chemical composition of the mosaic floor mortars. The chemical analyses were performed on finely crushed mortar samples by X-ray fluorescence spectroscopy (XRF) through a Panalytical Axios spectrometer PW4400/40 X-ray (Marvel Panalytical, Almelo, The Netherlands) equipped with Rh tube under argon/methane, at University of Aveiro (UA), using Omnian37 and Pro-Trace2021 software for major and minor elements analyses, respectively. The loss of ignition (LOI) was determined by heating the mortars samples at 1000 °C, using an electric furnace for 3 h. The major elements analyzed were Al_2O_3, CaO, Cl, Fe_2O_3, K_2O, MgO, MnO, Na_2O, P_2O_5, SiO_2, SO_3, and TiO_2 with detection limit of 1%. The detection limits of the trace elements analyzed were: As = 4.06 mg/kg, Ba = 6.90 mg/kg, Br = 0.78 mg/kg, Co = 4.54 mg/kg, Cr = 1.96 mg/kg, Cs = 4.78 mg/kg, Cu = 2.84 mg/kg, Ga = 0.94 mg/kg, Nb = 0.84 mg/kg, Ni = 2.00 mg/kg, Rb = 0.64 mg/kg, Sn = 3.02 mg/kg, Sr = 0.72 mg/kg, Th = 2.52 mg/kg, V = 2.78 mg/kg, Zn = 1.28 mg/kg and Zr = 0.80 mg/kg. Precision and accuracy of analyses and procedures were monitored using UA internal standards, certified reference material and quality control blanks. Results were within the 95% confidence limits. The relative standard deviation was between 5% and 10%.

All statistical analyses were performed using IBM SPSS® statistics v25. The normality of the data was verified (Shapiro–Wilk's test, $p > 0.05$). ANOVA, Tukey's test, t-student, K-means, cluster and discriminant analysis were used to determine groups and statistically significant differences ($p < 0.05$).

4. Results and Discussion

The chemical compositions, related to the binder and aggregates of the mosaic floor mortars samples, are detailed in Tables 2 and 3. The K-Means Cluster analysis of all samples identified two sets of samples in both major and trace elements, with MA1, MA4 and MD3 forming a cluster clearly separated from the remaining samples. The ANOVA 1-way analysis showed that the variables with significant statistical influence to define these two groups were Al_2O_3, K_2O, SiO_2, Ba, Cs, Ga, Nb, Rb, Sn ($p = 0.00$), MgO, Zr ($p = 0.001$), P_2O_5 ($p = 0.002$), Ni ($p = 0.003$), Br, Sr ($p = 0.006$), Na_2O ($p = 0.007$), and Fe_2O_3 ($p = 0.015$).

Table 2. Chemical concentrations of the major components of the mortars samples (in %).

Sample	Na_2O	MgO	Al_2O_3	SiO_2	P_2O_5	SO_3	Cl	K_2O	CaO	TiO_2	MnO	Fe_2O_3	LOI
MA2	0.34	1.10	30.93	35.42	2.98	0.24	0.13	2.60	5.11	1.25	0.06	8.38	11.24
MA3	0.30	1.16	32.49	37.07	1.50	0.40	0.10	2.68	3.65	1.30	0.07	8.48	10.62
MB1	0.27	1.35	32.68	38.53	3.45	0.86	0.04	3.43	2.14	1.35	0.13	8.62	6.93
MC1	0.33	0.97	29.27	31.87	4.64	3.82	0.11	2.46	5.48	1.12	0.10	7.58	11.96
MD1	0.43	1.11	33.02	32.90	4.57	0.41	0.19	2.76	3.85	1.27	0.11	8.63	10.53
MB2	0.18	1.28	34.97	36.16	3.46	0.35	0.03	2.86	1.52	1.37	0.09	8.91	8.59
MC2	0.26	0.99	34.09	34.63	4.35	0.32	0.10	2.52	2.41	1.22	0.17	8.09	10.63
MC3	0.22	1.14	34.03	35.65	3.87	0.26	0.03	3.35	1.46	1.62	0.10	11.07	6.84
MC4	0.19	1.35	35.93	36.42	3.50	0.17	0.07	2.70	1.70	1.28	0.16	8.50	7.78
MD2	0.43	1.47	34.08	34.91	3.39	0.23	0.15	3.26	2.41	1.38	0.08	8.69	9.22
MA4	1.11	2.25	22.27	47.31	0.74	0.21	0.07	6.70	5.60	1.37	0.10	7.01	4.83
MD3	1.74	1.70	20.71	54.06	1.98	0.07	0.03	7.07	2.01	1.22	0.07	6.22	2.76
MA1	0.23	1.56	26.11	47.56	0.86	0.13	0.03	4.73	1.11	1.54	0.13	7.80	7.77

Table 3. Chemical concentrations of the trace elements of the mortars samples (in mg/kg).

Sample	As	Ba	Br	Co	Cr	Cs	Cu	Ga	Nb	Ni	Pb	Rb	Sn	Sr	Th	V	Zn	Zr
MA2	23.1	350	60.2	5.6	45.4	18.6	23.7	36.0	31.4	11.6	59	270	28.4	85	36.4	50.4	77	300
MA3	17.3	360	52.4	5.1	47.4	22.2	27.0	37.1	32.3	13.3	51	250	23.2	65	37.0	56.1	87	300
MB1	51.9	520	30.6	8.4	46.4	20.8	67.2	34.5	29.2	13.8	71	290	21.0	160	42.2	67.9	170	340
MC1	55.6	420	67.4	8.6	40.1	20.6	55.8	33.3	29.9	11.2	240	240	21.9	87	34.2	60.6	78	250
MD1	24.9	440	85.8	8.4	50.0	18.6	57.5	37.3	32.2	14.3	56	250	23.6	90	37.9	57.3	190	280
MB2	76.6	380	35.1	8.8	51.8	19.8	55.7	38.1	33.5	13.7	61	260	24.4	70	40.9	63.6	120	310
MC2	58.5	330	65.0	15.2	45.6	18.9	71.4	36.1	32.2	12.4	170	250	23.6	67	36.2	69.6	130	270
MC3	61.5	560	38.1	8.0	58.3	22.2	58.9	44.7	37.6	14.6	660	290	29.6	88	48.1	87.5	130	400
MC4	45.7	370	42.1	15.6	48.4	20.5	62.5	38.0	34.3	13.2	77	250	25.1	59	36.8	68.0	200	300
MD2	25.3	440	65.4	4.9	48.2	19.9	64.0	36.1	31.3	15.0	53	270	21.3	89	39.2	65.7	190	340
MA4	4.9	950	20.3	7.3	32.9	17.5	35.3	21.2	18.6	7.8	44	380	10.5	210	37.8	62.5	170	450
MD3	14.4	970	12.5	<LOD	41.0	15.0	36.8	20.1	16.0	8.6	38	400	10.7	260	36.5	61.8	82	400
MA1	42.1	560	23.4	7.3	60.9	12.5	57.8	22.0	19.0	12.1	44	330	9.4	92	41.7	77.2	83	540

LOD: Limit of detection.

All samples were characterized by low CaO content (1.1–5.5%) and high concentration in F_2O_3 (6.2–11.1%) and Al_2O_3 (20.7–35.9%), when compared with the same elements from other studies on roman mortars [8,10,37]. A study on Pompeii mortars [38] found low percentages of CaO (3.81%) and attributed this low content to the high content of volcanic rock fragments as well as to the to the chemical composition of the binder composed by a mixture of lime and clay. Another possible explanation can be related to the degradation of those present study mortars (some cohesion loss was observed) as a result of the unfavorable humidity conditions to which the mosaic floor has been submitted that may lead to the leaching of the lime, as cohesion loss was observed during sampling [39]. The SiO_2 content was higher than Al_2O_3 in all samples, although in samples MA4 (between *nucleus* and *rudus*), MD3 (*rudus*) and MA1 (*statumen*), this difference was much more evident, with over 2X's. The highest SiO_2 content was found in samples MA4, MD3 and MA1, with 47.3%, 54.1% and 47.6%, respectively. These samples showed lower Al_2O_3 and higher K_2O content than nucleus ones (layers 1 and 2). MD3 showed the lowest Al_2O_3 (20.7%) and

LOI (2.7%) concentrations, among all samples. The low LOI (2.7–11.2%) in all samples, when compared to other studies, e.g., [8,40], can be related to the high percentage of clay minerals and its release of OH⁻ and calcite release of CO_2 [41].

Nucleus renders (layer 1—*nucleus* upper zone, and layer 2—*nucleus* lower zone) also displayed higher content of Fe_2O_3 (8.1–11.2%, except in sample MC1 with 7.6%), Al_2O_3 (29.3–36.0%) and lower SiO_2 content (31.8–38.5%), when compared to *rudus* (MD3) and *statumen* (MA1) render samples, which may be the result of ceramic powder presence in nucleus render composition. The higher content of Fe_2O_3 is responsible for the rose ochre color, that can be associated with the brick fragments or powder mixed in the mortar [38], and may indicate that iron is in its oxide form. Additionally, layer 1 showed higher CaO content (3.6–5.5%, except in sample MB1 with 2.1%) than layer 2 (1.5–2.4%). The higher CaO content in layer 1 may result from the influence of the composition of the bedding mortar that previously existed above the nucleus layer (although was not visible during sampling) and could be mixed in the nucleus upper zone. According to Vitruvius [33], it was composed by lime and very fine marble dust. In layer 1, MA2 and MC1 samples showed the highest CaO content, with 5.1% and 5.5%, respectively. The highest content in Al_2O_3 was observed in layer 2 (34.1–36.0%). The higher Cl content was found on the *nucleus* upper layer (layer 1), which may be the result of capillary rise with chloride soluble salts through the mosaic floor and evaporation occurring at the outer surface of the outer layer, and as a result higher Cl concentration was observed in this layer [42].

Previous studies about the chemical characterization of archaeological and ceramic materials used the hierarchical cluster analysis on chemical data to obtain a more robust interpretation [43,44]. The cluster analyses (Ward linkage) of layers 1 and 2, separated sample MC1 from the others, and defining two other subgroups, one with MA3, MB1, and MA2 samples, and a second one with MB2, MC4, MC3, MC2, MD2, and MD1 samples (Figure 8a). The variables cluster analysis also revealed that Al_2O_3 and SiO_2 present a distinct concentration pattern from the other major elements, in agreement with previous analysis (Figure 8b). An ANOVA 1-way analysis of these two groups revealed significant statistical differences in Al_2O_3 and CaO ($p < 0.05$). The groups descriptive statistics showed that Al_2O_3 concentration mean for layer 2 (36.62%) was higher than for layer 1 (31.68%), and that CaO mean for layer 1 was higher than for layer 2, with 4.04% and 1.90%, respectively. Despite differences identified, all the other variables did not present significant differences. Additionally, discriminant analysis based on the chemical composition of each nucleus layer (1 and 2) revealed that samples MC2 and MC4 from layer 2 showed more affinity with layer 1, and samples MA2 and MD1 from layer 1 revealed more affinity with layer 2. This may result from sampling contamination due to the small thickness of this layer. As this roman pavement is an archaeological heritage integrated in a museum, it is not possible to collect more samples.

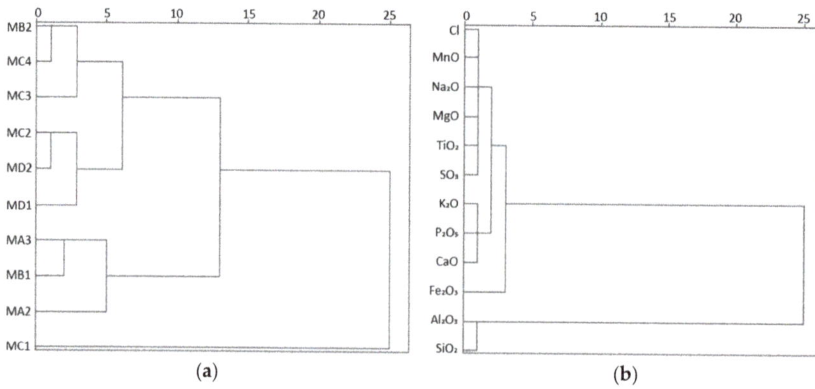

(a)   (b)

Figure 8. Cluster analysis of layer 1 and 2: (**a**) samples; (**b**) major elements.

Relatively high trace elements As, Pb and Cu contents (Table 3), when compared to other studies [38,40], were found. In most of the nucleus samples, the considerably higher As concentration might be attributed to the application of biocides to remove the biological colonization of the mosaic floor (e.g., [40]). The relatively high content in Cl, Pb and Cu (Table 2) can be due to the considerable degree of exposure to modern construction (concrete and Portland cement), considering that the mosaic floor is integrated in the new cement and concrete building of MDDS (e.g., [40]).

The dendrogram obtained by cluster analysis on the trace elements chemical data for all samples is presented in Figure 9. T-student test revealed significant differences between all variables in layers 1 and 2. An ANOVA 1-way analyses showed significant statistical differences between the 2 layers in V ($p < 0.05$) content. The descriptive statistics showed that the V mean in layer 2 (70.88 mg/kg) was higher than in group 1 (58.46 mg/kg). All the other variables did not present significant differences between groups. Moreover, discriminant analysis of layers 1 and 2 trace elements content suggested that samples MC2, MC3 and MC4 from layer 2 revealed more affinity with layer 1, and samples MC1 and MD1 from layer 1 showed more affinity with layer 2.

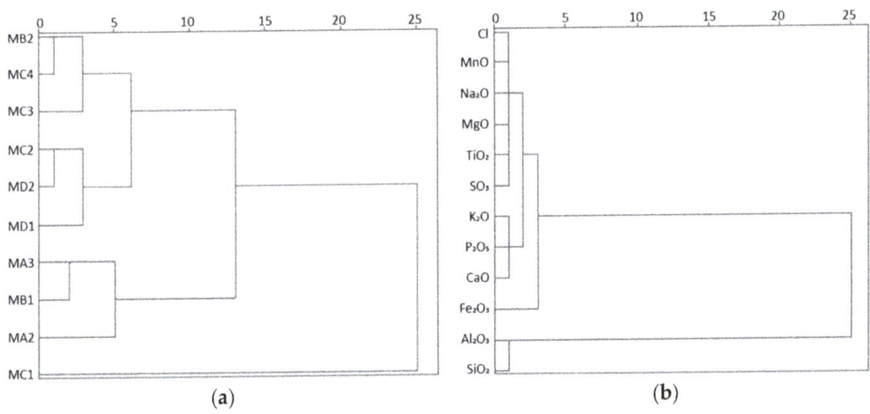

Figure 9. Dendogram obatined by cluster analyses: (**a**) samples; (**b**) trace elements—all groups.

These mortars were characterized by low content in CaO and high content of Fe_2O_3 and Al_2O_3. The *nucleus* (layers 1 and 2) was characterized by higher Fe_2O_3 and Al_2O_3 than *rudus* and *statumen*. *Rudus* and *statumen* showed higher SiO_2 content than *nucleus*. Comparing layers 1 with 2: layer 1 showed higher CaO content and layer 2 higher Al_2O_3 content. These layers showed relatively high As, Pb and Cu contents.

The statistical analyses enabled the differentiation of the two main *nucleus* layers (layers 1 and 2) from the three outlier samples: between *nucleus* and *rudus*, *rudus* and *statumen*. The descriptive statistics revealed differences among layers 1 and 2 concerning mainly Al_2O_3, CaO and V contents. Considering layers 1 and 2 major elements statistical analyses, MC2 and MC4 from layer 2 shows more affinity with layer 1 and MA2 and MD1 from group 1 shows more affinity with layer 2; analyses on trace elements revealed that samples MC1 and MD1 from layer 1 presented more affinity with layer 2 and MC2, MC3 and MC4 from layer 2 more affinity with layer 1.

5. Conclusions

The results obtained on the 13 mortars samples from different floor layers of the high imperial mosaics of the *domus* integrated in the crypt space of Museum of Archaeology D. Diogo de Sousa showed that there are differences on the chemical elements analyzed from the different layers: nucleus upper layer (layer 1), nucleus lower layer (layer 2), between *nucleus* (layer 3) and *rudus* (layer 4), *rudus* and *statumen* (layer 5) layers.

The chemical characterization of the studied floor mortar layers indicated that their composition was distinctly related to their stratigraphic position in the substrate, according to Vitruvius' model, and to the external conditions and treatments to which each layer was submitted (e.g., capillary rise with soluble salts and application of chemical treatments). Mosaic floor mortars showed low content of CaO and high F_2O_3, Al_2O_3, Cl, As, Pb and Cu contents. The statistical analyses using the chemical data on major and trace elements was robust and confirmed the clear separation between the *nucleus* layer (layers 1 and 2) and the other layers.

This study offers an important archaeological and historical contextualization and a first chemical characterization on the roman mortars of *Bracara Augusta*. Additionally, it is focused on mortar investigation of the oldest Roman housing testimonies of *Bracara Augusta* musealized in Museum of Archaeology D. Diogo de Sousa, Portugal.

Future studies will focus on the mineralogical analyses of the mosaic floor layers, and correlate them with the results of the chemical analyses discussed in this present study.

Author Contributions: Conceptualization, A.F., J.R. and F.R.; methodology, A.F. and F.R.; statistical analysis, C.C.; validation, A.F. and F.R.; formal analyses, A.F., J.R., C.C. and A.V. and F.R.; investigation, A.F., J.R.; writing—original draft preparation, A.F., J.R. and C.C.; writing—review and editing, A.F., J.R., C.C. and F.R.; supervision, A.F. and F.R. All authors have read and agreed to the published version of the manuscript.

Funding: This research was funded by GeoBioTec Research Centre (UIDB/04035/2020), funded by FCT—Fundação para a Ciência e a Tecnologia, FEDER funds through the Operational Program Competitiveness Factors COMPETE and by national funds (OE), through FCT, in the scope of the framework contract foreseen in the numbers 4, 5 and 6 of the article 23, of the Decree-Law 57/2016, of 29 August, changed by Law 57/2017, of 19 July.

Institutional Review Board Statement: Not applicable.

Informed Consent Statement: Not applicable.

Data Availability Statement: Not applicable.

Acknowledgments: The authors would like to thank the Museum of Archaeology D. Diogo de Sousa (MDDS), in Braga, Portugal, for sampling permission. We are very grateful to Vitor Hugo Torres from MDDS for his support and assistance in mortars sampling and to Cristina Sequeira for her assistance in mortars chemical analyses at Geosciences Department from University of Aveiro (UA).

Conflicts of Interest: The authors declare no conflict of interest.

References

1. Silva, A.S.; Paiva, M.; Ricardo, J.; Salta, M.M.; Monteiro, A.M.; Candeias, A.E. Characterisation of Roman mortars from the archaeological site of Tróia (Portugal). *Mater. Sci. Forum* **2006**, *514–516*, 1643–1647. [CrossRef]
2. Cardoso, I.; Macedo, M.F.; Vermeulen, F.; Corsi, C.; Santos Silva, A.; Rosado, L.; Candeias, A.; Mirao, J. A multidisciplinary approach to the study of archaeological mortars from the town of Ammaia in the Roman province of Lusitania (Portugal)*. *Archaeometry* **2014**, *56*, 1–24. [CrossRef]
3. Borsoi, G.; Santos Silva, A.; Menezes, P.; Candeias, A.; Mirão, J. Analytical characterization of ancient mortars from the archaeological Roman site of Pisões (Beja, Portugal). *Constr. Build Mater.* **2019**, *204*, 597–608. [CrossRef]
4. Velosa, A.L.; Coroado, J.; Veiga, M.R.; Rocha, F. Characterisation of roman mortars from Conímbriga with respect to their repair. *Mater. Charact.* **2007**, *58*, 1208–1216. [CrossRef]
5. Pachta, V.; Stefanidou, M. Technology of multilayer mortars applied in ancient floor mosaic substrates. *J. Archaeol. Sci.* **2018**, *20*, 683–691. [CrossRef]
6. Izzo, F.; Arizzi, A.; Cappelletti, P.; Cultrone, G.; De Bonis, A.; Germinario, C.; Graziano, F.; Grifa, C.; Guarino, V.; Mercurio, M.; et al. The art of building in the Roman period (89 B.C.–79 A.D.): Mortars, plasters and mosaic floors from ancient *Stabiae* (Naples, Italy). *Constr. Build. Mater.* **2016**, *117*, 129–143. [CrossRef]
7. Kramar, S.; Zalar, V.; Urosevic, M.; Körner, W.; Mauko, A.; BMirtič, B.; Lux, J.; Mladenović, A. Mineralogical and microstructural studies of mortars from the bath complex of the Roman *villa rustica* near Mošnje (Slovenia). *Mater. Charact.* **2011**, *62*, 1042–1057. [CrossRef]
8. Ergenç, D.; Fort, R. Multi-technical characterization of Roman mortars from Complutum, Spain. *Measurement* **2019**, *147*, 106876. [CrossRef]

9. Moreno-Alcaide, M.; Compaña-Prieto, J.M. Roman plasters and mortars from ancient Cosa (Tuscany-Italy). Mineralogical characterisation and construction from *domus* 10.1 (House with *Cryptoporticus*). *J. Archaeol. Sci.* **2018**, *19*, 127–137. [CrossRef]

10. Starinieri, V.; Papayianni, I.; Stefanidou, M. Study of materials and technology of ancient floor mosaics' substrate. *Conservar. Património* **2008**, *7*, 29–34. [CrossRef]

11. Miriello, D.; Barca, D.; Bloise, A.; Ciarallo, A.; Crisci, G.M.; De Rose, T.; Gattuso, C.; Gazineo, F.; La Russa, M.F. Characterisation of archaeological mortars from *Pompeii* (Campania, Italy) and identification of construction phases by compositional data analysis. *J. Archaeol. Sci.* **2010**, *37*, 2207–2223. [CrossRef]

12. Leone, G.; De Vita, A.; Magnani, A.; Rossi, C. Characterization of archaeological mortars from Herculaneum. *Thermochim. Acta* **2016**, *624*, 86–94. [CrossRef]

13. Martins, M.; Carvalho, H. Roman city of *Bracara Augusta* (Hispania Citerior Tarraconensis): Urbanism and territory occupation. *Agri Centuriate* **2017**, *14*, 79–98.

14. Martins, M. *Bracara Augusta*: A Roman town in the Atlantic area. In *Journal of Roman Archaeology*; JRA Supplementary Series 62; Thomson-Shore: Dexter, MI, USA, 2006; pp. 213–222.

15. Martins, M.; Braga, C.; Magalhães, F.; Ribeiro, J. Constructing identities within the periphery of the Roman Empire: North-west Hispania. In *Rome and Barbaricum. Contributions to the Archaeology and History of Interaction in European Protohistory*; Curcă, R.G., Rubel, A., Symonds, R., Voß, H.U., Eds.; Archaeopress: Oxford, UK, 2020; pp. 135–154.

16. Delgado, M.; Gaspar, A.M. Intervençâo arqueológica na Zona P1 (Antigas cavalariças do regimento de infantaria de Braga)—Archaeological intervention in zone P1 (Old stables of the Braga infantry regiment). *Cad. Arqueol.* **1986**, *3*, 155–167.

17. Martins, M.; Carvalho, H. *Bracara Augusta* and the changing rural landscape. In *Changing Landscapes. The Impact of Roman Towns in the Western Mediterranean, Proceedings of the International Colloquium, Castelo de Vide, Marvão, Portugal, 15–17 May 2008*; Corsi, C., Vermeulen, F., Eds.; Ante Quem: Bologna, Italy, 2010; pp. 281–298.

18. Martins, M.; Mar, R.; Ribeiro, J.; Magalhães, F. The Roman theatre of *Bracara Augusta*. In Proceedings of the XVIII CIAC: Centre and Periphery of the Ancient World, Mérida, Spain, 13–17 May 2013; pp. 861–864.

19. Magalhães, F. The Roman Domus on the NO Peninsula. Architecture, Construction and Sociability. Ph.D. Thesis, University of Minho, Braga, Portugal, 2019. (In Portuguese).

20. Delgado, M.; Martins, M. Intervençao arqueológica na Zona P1 (Antiguas cavalariças do regimento de infantaria de Braga)—Archaeological intervention in zone P1 (Old stables of the Braga Infantry Regiment). *Cad. Arqueol.* **1988**, *5*, 79–93.

21. Abraços, F.; Wrench, L. The Roman mosaics of *Bracara Augusta*: Re-reading and reinterpretation of decorative motifs. *J. Mosaic Res.* **2017**, *10*, 9–26. [CrossRef]

22. Abraços, F.; Lima, A.M.C.; Mourão, C.; Limão, F.; Wrench, L.; Pinto, M.M.; Gomes, P.D.; Lopes, R. *The Corpvs of the Roman Mosaics of Conventvs Bracaravgvstanvs*; Centro de Estudos Históricos: Lisbon, Portugal, 2019.

23. Martins, M.; Magalhães, F.; Peñín, R.; Ribeiro, J. The housing evolution of Braga between late Antiquity and the early Middle Ages. In *Arqueologia Medieval. Hàbitats Medievals*; Sabaté, F., Brufal, J., Eds.; Pagès Editors: Lleida, Spain, 2016; pp. 35–52.

24. Abraços, F. Conservation et restauration des mosaïques romaines au Portugal—Quelques exemples dans les collections de musées. In *Lessons Learned: Reflecting on the Theory and Practice of Mosaic Conservation; Proceedings of the 9th ICCM Conference, Hammamet, Tunisia, 29 November–3 December 2005*; Khader, B., Abed, A.B., Demas, M., Roby, T., Eds.; Getty Publications: Los Angeles, CA, USA, 2008; pp. 69–74.

25. Martins, M.; Carvalho, H. Transforming the territory: *Bracara Augusta* and its Roman cadaster. *Rev. Historiogr.* **2016**, *25*, 219–243.

26. Ferreira, N. *Geological Chart of Portugal in Scale 1:50,000—Sheet 5-D, Braga*; Ministério da Economia, Instituto Geológico Mineiro, Departamento de Geologia: Lisboa, Portugal, 2000.

27. Antunes, I.M.H.R.; Gonçalves, L.M.B. The "Sete Fontes" groundwater system (Braga, NW Portugal): Historical milestones and urban assessment. *Sustain. Water Resour. Manag.* **2018**, *5*, 235–248. [CrossRef]

28. Prudêncio, M.I. Ceramic in ancient societies: A role for nuclear methods of analysis. In *Nuclear Chemistry: New Research*; Koskinen, A.N., Ed.; Nova Science Publishers, Inc.: Hauppauge, NY, USA, 2008; Volume 2, pp. 51–81.

29. Ribeiro, J.; Martins, M. Os materiais de construção em *Bracara Augusta*. In *História da Construção—Os Materiais*; Melo, A., Ribeiro, M.V., Eds.; CITEM: Braga, Portugal, 2012; pp. 15–34.

30. Taelman, D.; Elburg, M.; Smet, I.; De Paepe, P.; Lopes, L.; Vanhaecke, F.; Vermeulen, F. Roman marble from Lusitania: Petrographic and geochemical characterisation. *J. Archaeol. Sci.* **2013**, *40*, 2227–2236. [CrossRef]

31. Ribeiro, J. *Arquitectura em* Bracara Augusta. *Uma análise das Técnicas Edilícias*; Afrontamento: Porto, Portugal, 2013.

32. Cazes, Q.; Areamond, J.C. Les fouilles du musée Saint Raymond à Toulouse (1994–1996). In *Mémoires de la Société Archeólogique du Midi de la France*; Conseil Général de La Haute-Garonne; Centre Atio Al D'études Spatiales: Toulouse, France, 1997; pp. 35–53.

33. Perrault, C. *Les Dix Livres d'Architecture de Vitruve*; Pierre Mardaga Editeur: Liège, France, 1996.

34. Fragata, A. Study of the mortar support of the mosaics of the imperial domus integrated in the space crypt of the regional museum of archaeology D. Diogo de Sousa (MDDS). In Proceedings of the XIV A.D. Saeculum Augustum—The Age of Augustus International Conference, Lisbon, Portugal, 24–26 September 2014. (In Portuguese).

35. Fragata, A. Compatible Renders for Masonries Submitted to Severe Water Action. Ph.D. Thesis, University of of Aveiro, Aveiro, Portugal, 2013. (In Portuguese).

36. Veiga, R.; Fragata, A.; Tavares, M.; Magalhães, A.; Ferreira, N. Inglesinhos convent: Compatible renders and other measures to mitigate water capillary rising problems. *J. Build. Apprais.* **2009**, *5*, 171–185. [CrossRef]

37. Alonso-Olazabal, A.; Ortega, L.A.; Zuluaga, M.C.; Ponce-Antón, G.; Jiménez Echevarría, J.; Fernández, A.C. Compositional characterization and chronology of roman mortars from the archaeological site of Arroyo de la Dehesa de Velasco (Burgo de Osma—Ciudad de Osma, Soria, Spain). *Minerals* **2020**, *10*, 393. [CrossRef]
38. Miriello, D.; Bloise, A.; Crisci, G.M.; De Luca, R.; De Nigris, B.; Martellone, A.; Osanna, M.; Pace, R.; Pecci, A.; Ruggieri, N. New compositional data on ancient mortars and plasters from *Pompeii* (Campania—Southern Italy): Archaeometric results and considerations about their time evolution. *Mater. Charact.* **2018**, *146*, 189–203. [CrossRef]
39. Lubelli, B.; Nijland, T.; Hees, R. Self-healing of lime based mortars: Microscopy observations on case studies. *Heron* **2011**, *56*, 75–91.
40. Ontiveros-Ortega, E.; Rodríguez-Gutiérrez, O.; Navarro, A. Mineralogical and physical-chemical characterization of Roman mortars used for monumental substrates on the hill of San Antonio, in the Roman city of Italica (prov. *Baetica*, Santiponce, Seville, Spain). *J. Archaeol. Sci. Rep.* **2016**, *7*, 205–223.
41. Daoudi, L.; Rocha, F.; Costa, C.; Arrebei, N.; Fagel, N. Characterization of rammed earth materials from the XVIth century. Badii palace in Marrakech, Morocco to ensure authentic and reliable restoration. *Geoarchaeology* **2017**, *33*, 529–541. [CrossRef]
42. Fragata, A.; Veiga, R.; Velosa, A. Substitution ventilated render systems for historic masonry: Salt crystallization tests evaluation. *Const. Build. Mater.* **2016**, *102*, 592–600. [CrossRef]
43. Liritzis, I.; Xanthopoulou, V.; Palamara, E.; Papageorgiou, I.; Iliopoulos, I.; Zacharias, N.; Vafiadou, A.; Karydas, A. Characterization and provenance of ceramic artifacts and local clays from late Mycenaean Kastrouli (Greece) by means of p-XRF screening and statistical analysis. *J. Cult. Herit.* **2020**, *46*, 61–81. [CrossRef]
44. Papageorgiou, I. Ceramic investigation: How to perform statistical analyses. *Archaeol. Anthrop. Sci.* **2020**, *12*, 210. [CrossRef]

applied
sciences

MDPI

Article

MA-XRF for the Characterisation of the Painting Materials and Technique of the *Entombment of Christ* by Rogier van der Weyden

Anna Mazzinghi [1,2,*], Chiara Ruberto [1,2], Lisa Castelli [2], Caroline Czelusniak [2], Lorenzo Giuntini [1,2], Pier Andrea Mandò [1,2] and Francesco Taccetti [2]

[1] Department of Physics and Astronomy, Università Degli Studi di Firenze, 50019 Sesto Fiorentino, Italy; ruberto@fi.infn.it (C.R.); giuntini@fi.infn.it (L.G.); mando@fi.infn.it (P.A.M.)
[2] Florence Division, INFN/CHNet, 50019 Sesto Fiorentino, Italy; castelli@fi.infn.it (L.C.); czelusniak@fi.infn.it (C.C.); ftaccetti@fi.infn.it (F.T.)
* Correspondence: mazzinghi@fi.infn.it

Abstract: At present, macro X-ray fluorescence (MA-XRF) is one of the most essential analytical methods exploited by heritage science. By providing spatial distribution elemental maps, not only does it allow for material characterisation but also to understand, or at least to have a likely idea of, the production techniques of an analysed object. INFN-CHNet, the Cultural Heritage Network of the Italian National Institute of Nuclear Physics, designed and developed a MA-XRF scanner aiming to be a lightweight, easy to transport piece of equipment for use in in situ measurements. In this study, the INFN-CHNet MA-XRF scanner was employed for the analysis of a painting by the Flemish artist Rogier van der Weyden. The painting belongs to the collection of the Uffizi gallery in Florence and was analysed during conservation treatments at the Opificio delle Pietre Dure, one of the main conservation centres in Italy. The research aims were to characterise the materials employed by the artist and to possibly understand his painting technique. Although MA-XRF alone cannot provide a comprehensive characterisation, it nonetheless proved to be an invaluable tool for providing an initial overview or hypothesis of the painting materials and techniques used.

Keywords: MA-XRF; heritage science; non-invasive analysis; portable equipment; pigment identification; van der Weyden; Flemish painting; calco-potassic glass in painting; INFN-CHNet; Opificio delle Pietre Dure

Citation: Mazzinghi, A.; Ruberto, C.; Castelli, L.; Czelusniak, C.; Giuntini, L.; Mandò, P.A.; Taccetti, F. MA-XRF for the Characterisation of the Painting Materials and Technique of the *Entombment of Christ* by Rogier van der Weyden. *Appl. Sci.* **2021**, *11*, 6151. https://doi.org/10.3390/app11136151

Academic Editor: Letizia Bonizzoni

Received: 5 June 2021
Accepted: 28 June 2021
Published: 2 July 2021

Publisher's Note: MDPI stays neutral with regard to jurisdictional claims in published maps and institutional affiliations.

1. Introduction

Macro X-ray fluorescence mapping, widely known as MA-XRF in the scientific community [1], is now one of the most valuable tools aiding the heritage science field. MA-XRF analysis produces elemental distribution maps of a scanned area of an object, proving to be exceptionally useful not only for material characterisation but also for understanding an artist's painting and production techniques. In addition, mapping makes it easier to associate trace elements with a specific colour or area, even in noisy maps, with respect to a "traditional" single point measurement. Moreover, as MA-XRF renders not only spectra but also images, it can also be easily adopted by non-X-ray experts, such as conservators or art historians.

The technique is relatively fast; the acquisition time naturally depends on the dimensions of the area of analysis and the dimensions and characteristics of the instrument employed. Numerous types of equipment are available nowadays, ranging from small and easy-to-transport instruments customized for the analysis of areas of only a few cm^2 [2,3] to massive scanners allowing large areas to be analysed in a reasonably short acquisition time [4–6].

Numerous successful applications of MA-XRF in heritage science have now been described in many papers. Many of these applications involve the analysis of paintings [7–10], but these applications also range from manuscripts [11,12], painted papers [13], metals [14], glasses [15], tiles [16], and archaeological sites [17] to furniture [18] and many others.

The research presented here used the MA-XRF scanner specifically designed and developed for heritage science by INFN-CHNet, the Cultural Heritage Network of the Italian National Institute of Nuclear Physics [19–21]. There are several partners within this network, one of these being the Opificio delle Pietre Dure (OPD) in Florence, a reputable conservation centre belonging to the Italian Ministry of Cultural Heritage. The Florence division of INFN works in close and fruitful collaboration with OPD, mainly for the analysis of paintings. One of these is the focus of this work, a painting produced by the famous Flemish painter known as Rogier van der Weyden, who was active in the first half of the XV century. This painting, the *Entombment of Christ*, belongs to the collection of the Uffizi gallery in Florence and was likely painted in 1460. The main aim of this work was to gain further information on the painting palette and techniques used by van der Weyden, comparing these results with those from other works by the artist, as specifically described in the results section. The materials analysis also aimed at supporting the ongoing conservation intervention.

To accomplish this purpose, the INFN-CHNet MA-XRF scanner has been proven to be an invaluable tool for an early non-invasive analysis of the elemental composition of the materials, the most of which could be reasonably hypothesised. Needless to say, MA-XRF alone cannot conclusively identify most of the pigments or other materials and thus cannot ascertain their exact nature. Other analytical methods are indeed routinely employed in combination with a comprehensive material characterisation. It is, however, true that MA-XRF analysis allows for a relatively fast and complete analysis able to provide, at the very least, a first overview or hypothesis of the materials and techniques employed by the artist, as will be demonstrated in this work. These results can be considered a first reliable basis for material identification and may act as an guide for the use of consecutive analytical methods.

2. Materials and Methods

2.1. The Entombment of Christ by Rogier van der Weyden

Rogier van der Weyden, also known as Rogier de la Pasture, is considered one of the most important and influential Flemish painters of his time, among contemporaries such as Jan van Eyck. He was born in Tournai in 1399 and settled in Brussels during 1435, becoming the official painter of the city in just three years, where he died in 1464 [22]. Both masters were innovative, primarily for their use of multiple thin transparent oil layers that achieved unprecedented colour intensities. The technique also gave them the ability to carefully paint fine and minute details, as can be seen, for example, in detailed decorations of precious textiles or in the abundance of naturalistic details of extremely realistic-looking scenes. Rogier van der Weyden not only demonstrated his technical skills in naturalistic detail, but also added emotions as an important element of the naturalism of characters, painting people moved to tears and covering their faces to express their grief or sorrow. These features are clearly visible in the *Entombment of Christ* from the Uffizi (Figure 1), which was possibly commissioned by the Medici family for their Villa at Careggi, near to Florence [23]. Moreover, it is likely that the Medici were one of the artist's Italian clients [22]. Indeed, the fame of the Netherlandish artist had spread throughout Europe by the middle of the 15th century, and he also travelled to Italy in 1450 where he most likely encountered Italian artists and works of art. It is therefore realistic that his growing fame attracted an increasing demand for his work and that he needed the aid of several skilled assistants [24]. This is one of the reasons why works by van der Weyden are rarely securely identified as his own, as he had a very large productive workshop in Brussels, and in those times works were not signed by artists. It is thus often difficult to separate the master's unaided work from that of his assistants and imitators [25].

Figure 1. The *Entombment of Christ* by Rogier van der Weyden.

This painting represents the burial of Christ with the Virgin and Saint John the Evangelist holding his hands (Figure 1). The body is supported by Joseph of Arimathea (an old man at Christ's proper right) and Nicodemus (the man dressed in green at Christ's proper left); kneeling Magdalene is in the lower foreground. The scene takes place in front of the tomb located on a rocky and hilly landscape with a city in the distance. Two Marys are on their way to the tomb in the left side of the background. The scheme of the scene is possibly inspired by the predella painted by Beato Angelico [23] depicting the Deposition of Christ, a painting that van der Weyden most likely saw in Florence during his travel in Italy.

The painting (oil on panel) is in pristine condition, and the most recent conservation intervention at the OPD was aimed mainly at a superficial treatment in the occasion of an exhibition at the Mauritshuis in 2018 [26].

2.2. The INFN-CHNet MA-XRF Scanner

The instrument was designed with a special focus on portability and lightness. The technical characteristics and analytical capabilities (detection efficiency, spatial resolution, etc.) of this equipment are thoroughly described in [27]. Briefly, the measuring head of the instrument was composed of an X-ray tube (Moxtek, 40 kV maximum voltage, 0.1 mA maximum anode current) and a Silicon Drift Detector (Amptek XR100 SDD, 25 mm^2 effective active surface, 500 μm thickness). A telemeter (Keyence IA-100) that was also installed on the measuring head, acted for the control and adjustment of the sample-instrument distance during the scan. This measuring head was mounted on three linear motor stages by Physik Instrumente, with a 200 mm travel range in the x and y directions for this version, plus a 50 mm stage along the z perpendicular direction. This was all installed on a carbon-fibre box containing motor controllers, a signal digitizer (CAEN DT5780), and other auxiliary elements. The software controlling acquisition and data analysis was developed by INFN-CHNet. The instrument has been successfully employed in several heritage science applications over the years [28–30].

During the analysis discussed here, the operating conditions of the X-ray tube for all measurements were: 30 kV anode voltage, 0.1 mA filament current, and a Mo anode with an 800 μm diameter collimator. The scanning velocity ranged from 1 to 2 mm/s and the equivalent pixel size ranged from 200 to 1000 μm. Typical measuring conditions, which are those used in the maps shown in this paper, were a scanning speed of 1 mm/s and a pixel size of 1 mm. Different parameters are specified when needed in the Supplementary Materials.

3. Results and Discussion

In the present study, several areas were analysed to investigate the materials employed in this painting. The materials identified/hypothesised are broadly consistent with those that were available to artists in the second half of the 15th century. The most interesting areas of analysis and corresponding XRF maps will be shown and discussed in detail in this paragraph. Other areas and maps can be found in the Supplementary Materials.

Pb is present almost ubiquitously in the painting, or at least in the areas analysed, most likely in the form of lead white used in mixtures with other pigments and dyes to achieve lighter hues. Lead white was possibly employed also in the imprimitura layer, as happened commonly in paintings during those times and frequently in Flemish paintings after the second quarter of the fifteenth century (as consistent with the dating of the painting) [31]. Furthermore, calcite traces in lead white were found by de Viguerie et al. [32] in other works by Rogier van der Weyden. In this painting, the presence of Ca associated to lead white could not be conclusively established as, for different reasons as will be further discussed, Ca is likely present also in other pigments/colours.

Blue areas, such as those of the Virgin's and Magdalene's robes—the former with intense hue and the latter in a pale tone—or the blue damask decorations of Nicodemus (see discussed later and Figure S1, at higher spatial resolution), are characterised by the presence of Cu, most likely indicating the use of azurite, rather than other Cu-based blue pigments such as blue verditer or Egyptian blue, as found in other works by the artist (Figure 2) [33]. Zn traces associated with these Cu-based areas are consistent, though not common [34], with traces detected in some cases in azurite, corroborating the hypothesis [35] (see Figure S2). Possibly, the few X-ray counts of Ca detected in these blue areas might also be related to calcite traces in azurite [36] (see Figure S1), though it cannot be specifically excluded that the Ca detected is also due to the gypsum-based preparation layer.

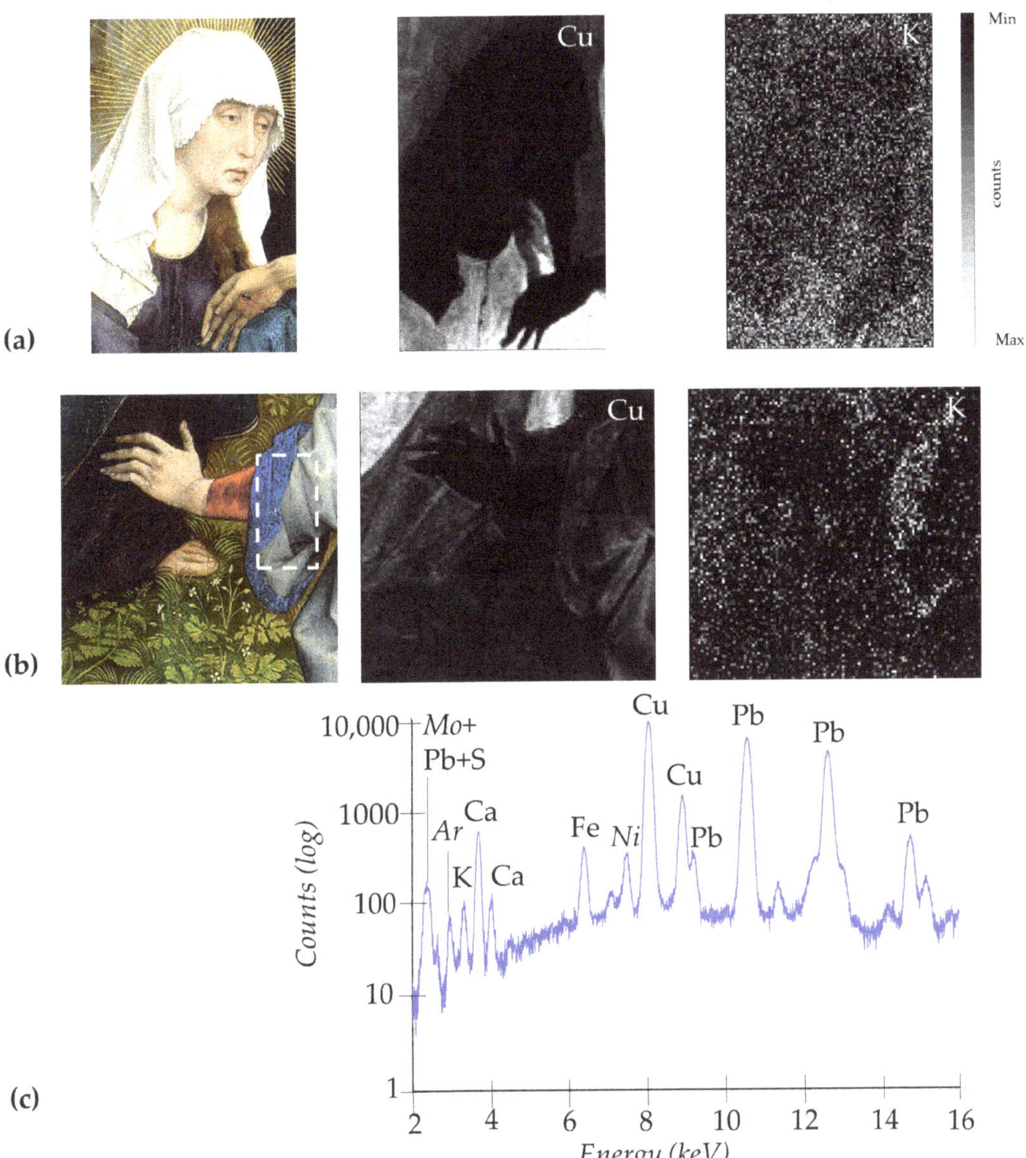

Figure 2. (a) Detail of the blue and purple Virgin's robe, with relative Cu(Kα) and K(K) XRF maps; (b) detail of Magdalene's robe (with purple and blue Virgin's robe on the back), with relative Cu(Kα) and K(K) XRF maps; (c) spectrum of the blue sleeve of Mary Magdalene (relative to the dashed area in detail (b)). Detected Ni is due to impurities in the X-ray tube, Mo is due to the X-ray tube anode, and Ar is due to the air path.

In those times, it was common to use an azurite underpaint layer beneath a more expensive ultramarine thin layer [37], a practice also attested to other works by Rogier van der Weyden [33,38]. The distribution of K maps, though present in traces, may suggest that this layering also occurs in the *Entombment* from the Uffizi (Figure 2). Indeed, K, though not specifically present in the blue mineral lazurite principal component of ultramarine, is characteristic of this precious pigment, as it takes part in the production process [39,40].

Cu-based compounds, possibly azurite anew, were also contained in the deep purple robe of the Virgin (Figure 2) and in the reddish/amaranth mantle of Joseph of Arimathea (Figure 3). Most likely, azurite was mixed with variable amounts of organic dye(s) to obtain these different hues, as might be hypothesised by the detection of K traces in the purple robe of the Virgin (Figure 2), characteristic of the alum commonly employed in the production process of organic dyes and lakes [41]. At the same time, it seems unlikely that in this case K indicates the use of the expensive ultramarine in such a hue, although it cannot be completely excluded. Iron oxides/hydroxides compounds (possibly earth or ochres) are also present in small amounts (Figure 3). Instead, no Hg was detected in these areas, however it was used in the fur trimming of Joseph of Arimathea's mantle together with the Fe-based compounds and lead white (Figure 3).

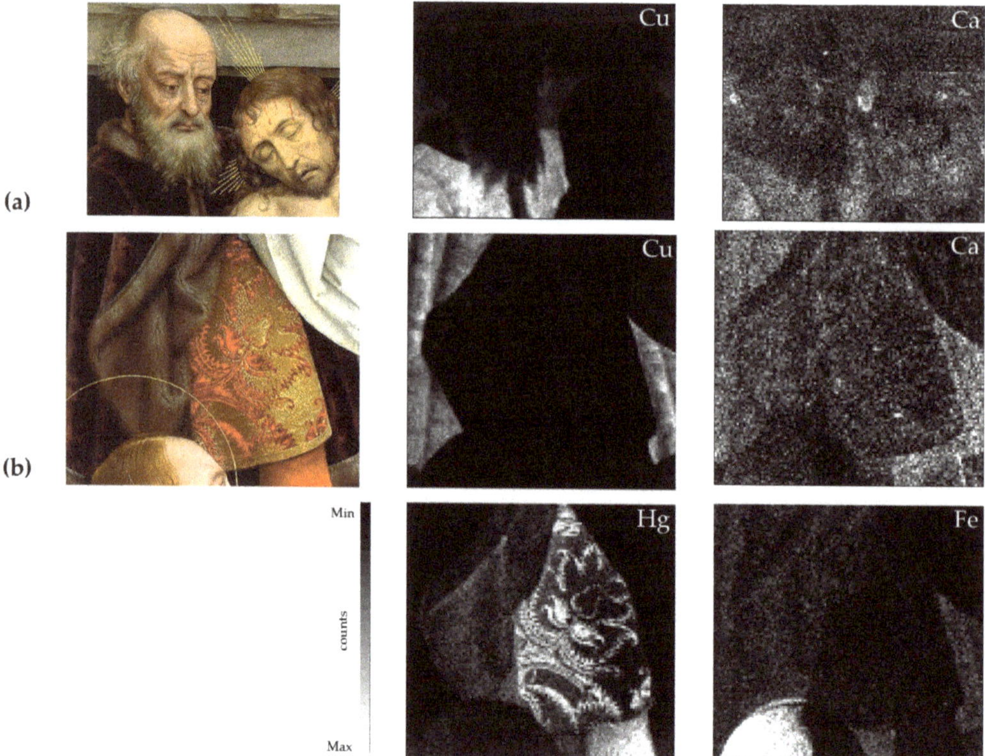

Figure 3. (a) Detail of the red/amaranth robe of Joseph of Arimathea, with relative Cu(Kα) and Ca(Kα) XRF maps; (b) detail of the red/amaranth robe of Joseph of Arimathea (lower part), with relative Cu(Kα), Fe(Kα), Hg(Lα), and Ca(Kα) XRF maps.

Ca was also detected in association with these purple/reddish areas (Figure 3) and might be related to the gypsum preparation layer, as discussed earlier. However, it cannot be excluded that it might also be due to the use of glass as an additive in the painting layer. Powdered glass, acting most likely as a drier/siccative of the oil medium, was indeed used extensively by artists, particularly between the fifteenth and sixteenth centuries in Italy and in Northern European works [42–44]. It was particularly suitable for red lakes, as it does not affect the typical transparency of these materials. In Netherlandish paintings, calco-potassic glasses were common, in which the main constituents (other than silica) are variable amounts of Ca and K due to ashes and/or lime used in the production process [45], with lower amounts of Mn in comparison to glasses used in Italian paintings [46]. Interestingly, it is documented that finely ground Ca-rich glasses were also used as additives by van der Weyden [33].

The red garment of John the Evangelist, together with other small red areas of the robe of one of the two Marys in the left side of the background landscape, are characterised by the presence of Hg, indicating the use of vermilion (Figure 4) as well as red damask decorations on Joseph of Arimathea's robe (Figures 3 and S1, at higher spatial resolution). Modelling was obtained by means of Cu- and Fe oxides/hydroxide-based compounds (both present in low quantities), as similarly happens in other garments (Figure 4). The red roof of the building on the right side of the background was painted with a mixture, or overlay, of vermilion with Fe-based compounds (See Figure S3). Vermilion areas are associated with Zn traces, as is commonly attested to this pigment [47] (see Supplementary Materials S4). It is worth noting that Ca was detected in association with red draperies and robes, especially in the deepest shadows, and is well visible in the red robe of Saint John the Evangelist and in the red sleeve of Magdalene (Figures 4 and S4). The presence of Ca in these areas may suggest anew, as previously discussed, the possible presence of Ca-rich glass as an additive in the red lakes used in the modelling of red draperies, as also attested in other works by van der Weyden [25].

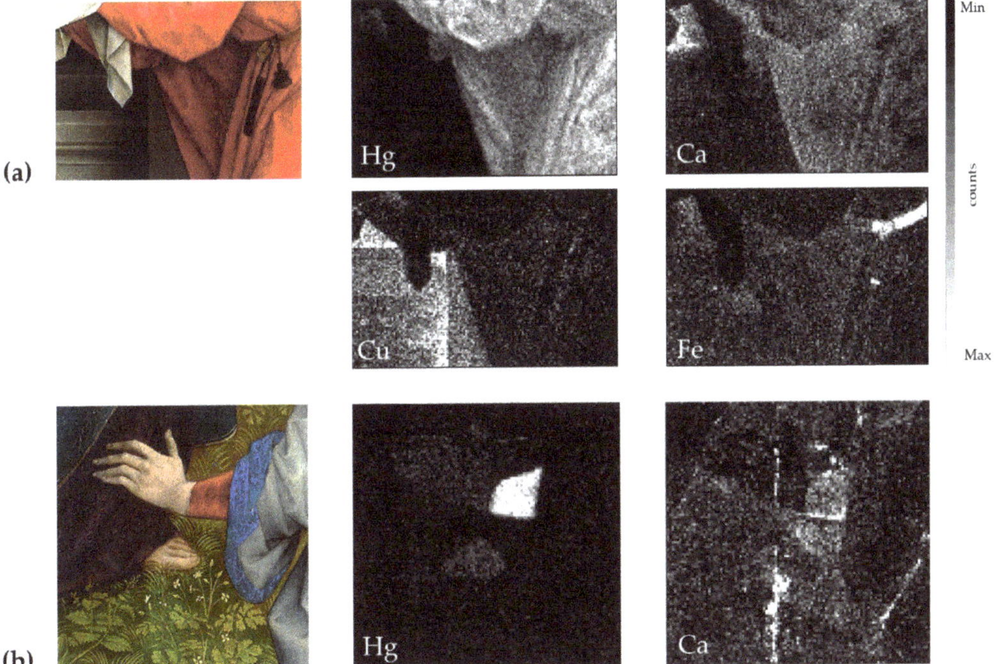

Figure 4. (**a**) Detail of the red robe of John the Evangelist, with relative Hg(Lα), Ca(Kα), Cu(Kα), and Fe(Kα) XRF maps; (**b**) detail of the red sleeve of Magdalene, with relative Hg(Lα) and Ca(Kα) XRF maps.

Cu is also characteristic of the green garment of Nicodemus (Figure 5). A wide range of Cu-based greens was available in those times and cannot be conclusively identified by means of XRF alone. It is worth noting, however, that verdigris and copper sulphates were detected in other paintings by van der Weyden [38,42]. Sn was also detected in association with this Cu-based green robe, attesting the likely use of lead-tin yellow in modelling

(Figure 5). Lead-tin yellow was also employed in the golden damask robes of Nicodemus and Joseph of Arimathea, where vermilion and Fe-based compounds were employed for modelling (Figure 5, showing Nicodemus' robe). Green vegetation, grass, and trees were painted with Cu-based compounds and lead–tin yellow as well, not necessarily in mixture but possibly employed in layers to build the volumes of the landscape (Figure 5). Small amounts of Fe-based compounds were also detected, likely not in association with K and thus in principle excluding the use of green earth. Ca was also detected in some cases in association with trees, a presence that could not be conclusively understood (see Figure S5). It can be hypothesised that the artist made the use of lakes (with powdered glass anew) or of ivory/bone black, though its presence is difficult to confirm with XRF alone, as will be further discussed.

Cu-based pigments and vermilion were also employed in light blue/grey and pink background architectures, respectively (see Figure S3).

Flesh tones were painted using lead white and vermilion with details and outlines obtained with Fe-based pigments. Cu-based compounds were employed, in small amounts, in the shadows and eyes (Figures 5, 6, S2 and S6; the latter at higher spatial resolution where the use of vermilion was highlighted in tiny detail such as the red lacrimal caruncle of the eye). Hair was obtained with a complex mixture of vermilion, Fe- and Cu- based compounds, as well as lead-tin yellow (Figures 6, S2 and S5), where different hair tones were obtained with various mixtures of these pigments. In grey hair, such as that of Joseph of Arimathea, mainly Cu-based pigments and lead white were employed, together with small quantities of lead-tin yellow (Figure 6). The blood drops on Christ's forehead were painted with vermilion (Figure 6).

Golden lines representing the halos or rays around the character's heads were realised with gold (Figures 6 and S2). Any distribution of other elements leads to the thought of the presence of a preparation layer, though the golden lines are so thin that associations with their distribution could not be properly evaluated. Upon visual inspection, gildings were likely applied using the mission technique.

The rocks and cave of the burial were painted in grey/dark tones. Cu is characteristic of these areas (Figure 6) together with Zn traces, as previously discussed for the blue areas. Darker tones of the cave entrance were painted with the evident addition of Fe-based pigments (Figures 5 and 6). The black hat of Nicodemus was painted with a similar mixture, with the addition of a significant amount of Ca (Figure 5), which was also detected in the darker zones of the cave (Figure 5). P could not be detected in these areas; thus, it seems unlikely that van der Weyden used bone/ivory black, although its presence cannot be conclusively excluded. P XRF detection efficiency is indeed rather low, especially on varnished paintings. In addition, there was a partial overlap of the escape peak of the Ca Kα X-ray line with that of P, further encumbering its detection. It is also possible that the artist added some lakes, with Ca-rich glass, to warm this black hue.

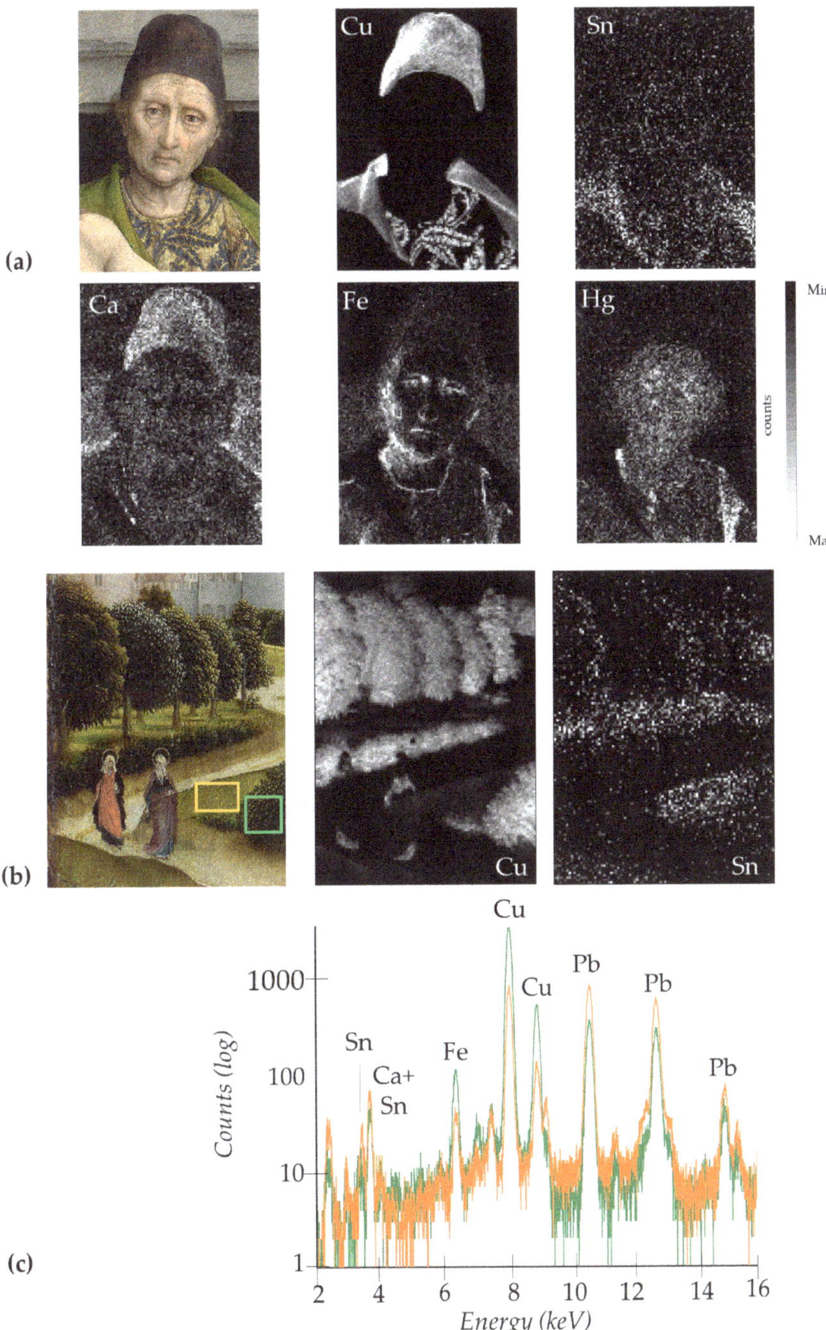

Figure 5. (**a**) Detail of Nicodemus, with relative Cu(Kα), Sn(Lα), Ca (Kα), Fe(Kα), and Hg(Lα) XRF maps; (**b**) detail of the upper left landscape, with relative Cu(Kα) and Sn(Lα) XRF maps; (**c**) spectra of the green and yellowish landscape, relative to the areas indicated in detail (**b**). The two ROI are roughly of the same area in pixels.

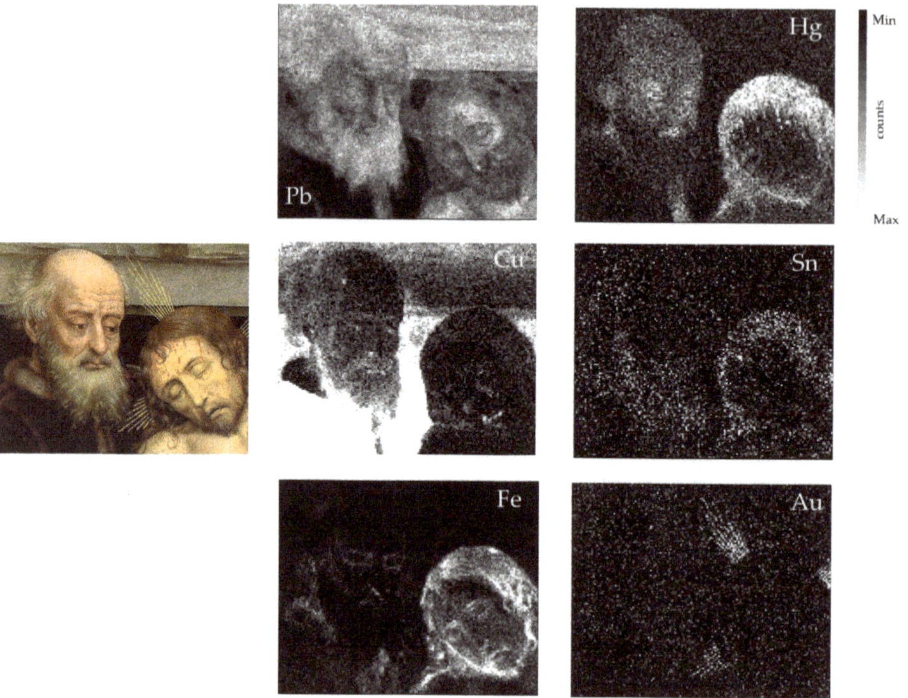

Figure 6. Detail of Joseph of Arimathea and Christ, with relative Pb(Lα), Hg(Lα), Fe(Kα), Cu(Kα), Sn(Lα), and Au (Lα) XRF maps. The grey scale in the Cu map is edited to enhance the distribution in flesh tones and hair.

4. Conclusions

The imaging approach of MA-XRF employed in this study allowed to obtain information regarding the materials and techniques used by Rogier van der Weyden in his *Entombment of Christ*. With a degree of uncertainty due to the technique itself, traditional materials were determined or hypothesised by XRF. These include azurite, ultramarine, lead–white, vermilion, Cu-based green, Fe oxides/hydroxides (earth/ochres), lead–tin yellow, and gold (in thin lines in halos). The detection of Ca traces associated with some areas possibly suggests the use of dyes or lakes in which a calco-potassic glass, present as a drier/siccative, is consistent with the Flemish technique during those times. Although committed by the Medici's family, this result may also in principle confirm the Flemish production of the painting and exclude the possibility that it was painted in Florence during the artists' journey to Italy, as the Italian glass used for this purpose were Mn-rich rather than Ca-rich.

As it has been demonstrated also in this study, MA-XRF allows the non-invasive mapping of the distribution of the different painting materials in sufficient detail, but at the same time it is true that unfortunately it lacks the specificity for a comprehensive characterisation of all materials. It is indeed highly recommended to combine, when available, other analytical methods for an accurate identification of the compounds. However, MA-XRF is undoubtedly a powerful technique which is a source of relevant information, especially when employed as an early non-invasive and non-destructive analytical method acting as a guide for a subsequent more accurate scientific analysis.

Supplementary Materials: The following are available online at https://www.mdpi.com/2076-3417/11/13/6151/s1: Figure S1: (a) Detail of the damask robe of Joseph of Arimathea, with relative Hg(Lα) and Pb(Lα) XRF maps; (b) Detail of the damask robe of Nicodemus, with relative Cu(Lα),

K(Kα), Hg(Lα), Sn(Lα) and Ca(Kα) XRF maps. Measuring conditions (both areas): scanning speed 2 mm/s, pixel size 250 μm. Figure S2: Detail of the Virgin, with relative Pb(Lα), Fe(Kα), Au(Lα), Hg(Lα), and Zn(Kα) XRF maps. Figure S3: Detail of the city in the far right of the background, with relative Hg(Lα), Fe(Kα), Ca(Kα), Cu(Kα), and Sn(Lα) XRF maps. Figure S4: Detail of Magdalene, with relative Pb(Lα), Fe(Kα), Hg(Lα), Sn(Lα), Zn(Kα) and Ca(Kα) XRF maps. Figure S5: Detail of the upper left landscape, with relative Ca(Lα), Fe(Kα), Pb(Lα) and Zn(Kα) XRF maps. Figure S6: Detail of the eyes of John the Evangelist, with relative Fe(Kα), Hg(Kα), Pb(Lα) and Cu(Kα) XRF maps. Measuring conditions: scanning speed 2 mm/s, pixel size 250 μm.

Author Contributions: Conceptualization, A.M., C.R., L.C.; methodology, F.T., L.G., P.A.M.; software, F.T., C.C., L.C.; validation, P.A.M.; formal analysis, A.M., C.R.; investigation, A.M., C.R.; resources, F.T., L.G., P.A.M.; data curation, F.T., C.C., L.C.; writing—original draft preparation, A.M.; writing—review and editing, A.M., L.C., C.R., L.G.; visualization, A.M., C.R., L.C.; supervision, F.T., P.A.M.; project administration, F.T. All authors have read and agreed to the published version of the manuscript.

Funding: This research received no external funding.

Acknowledgments: The authors are grateful to the *Opificio delle Pietre Dure* and the *Gallerie degli Uffizi* for allowing this study. In particular, Cecilia Frosinini, Roberto Bellucci, Chiara Rossi, Chiara Modesti (OPD), Eike D. Schmidt and Carolina Forasassi (Uffizi) are gratefully acknowledged. Authors warmly thank Lara Palla for her contribution to the MA-XRF software. Thanks are also due to Marco Manetti for his invaluable technical support and to Leandro Sottili for his helpful suggestions.

Conflicts of Interest: The authors declare no conflict of interest.

References

1. Alfeld, M. MA-XRF for Historical Paintings: State of the Art and Perspective. *Microsc. Microanal.* **2020**, *26*, 72–75. [CrossRef]
2. Galli, A.; Caccia, M.; Alberti, R.; Bonizzoni, L.; Aresi, N.; Frizzi, T.; Bombelli, L.; Gironda, M.; Martini, M. Discovering the material palette of the artist: A p-XRF stratigraphic study of the Giotto panel 'God the Father with Angels'. *X-ray Spectrom.* **2017**, *46*, 435–441. [CrossRef]
3. Alfeld, M.; Mulliez, M.; Martinez, P.; Cain, K.; Jockey, P.; Walter, P. The Eye of the Medusa: XRF Imaging Reveals Unknown Traces of Antique Polychromy. *Anal. Chem.* **2017**, *89*, 1493–1500. [CrossRef] [PubMed]
4. Alfeld, M.; Pedroso, J.V.; Hommes, M.V.E.; Van der Snickt, G.; Tauber, G.; Blaas, J.; Haschke, M.; Erler, K.; Dik, J.; Janssens, K. A mobile instrument for in situ scanning macro-XRF investigation of historical paintings. *J. Anal. At. Spectrom.* **2013**, *28*, 760–767. [CrossRef]
5. Romano, F.P.; Caliri, C.; Nicotra, P.; Di Martino, S.; Pappalardo, L.; Rizzo, F.; Santos, H.C. Real-time elemental imaging of large dimension paintings with a novel mobile macro X-ray fluorescence (MA-XRF) scanning technique. *J. Anal. At. Spectrom.* **2017**, *32*, 773–781. [CrossRef]
6. Pouyet, E.; Barbi, N.; Chopp, H.; Healy, O.; Katsaggelos, A.; Moak, S.; Mott, R.; Vermeulen, M.; Walton, M. Development of a highly mobile and versatile large MA-XRF scanner for in situ analyses of painted work of arts. *X-ray Spectrom.* **2021**, *50*, 263–271. [CrossRef]
7. Saverwyns, S.; Currie, C.; Delgado, E.L. Macro X-ray fluorescence scanning (MA-XRF) as tool in the authentication of paintings. *Microchem. J.* **2018**, *137*, 139–147. [CrossRef]
8. Reiche, I.; Eveno, M.; Müller, K.; Calligaro, T.; Pichon, L.; Laval, E.; Mysak, E.; Mottin, B. New insights into the painting stratigraphy of L'Homme blessé by Gustave Courbet combining scanning macro-XRF and confocal micro-XRF. *Appl. Phys. A* **2016**, *122*, 947. [CrossRef]
9. Ravaud, E.; Pichon, L.; Laval, E.; Gonzalez, V.; Eveno, M.; Calligaro, T. Development of a versatile XRF scanner for the elemental imaging of paintworks. *Appl. Phys. A* **2016**, *122*, 1–7. [CrossRef]
10. Delaney, J.K.; Dooley, K.A.; Van Loon, A.; Vandivere, A. Mapping the pigment distribution of Vermeer's Girl with a Pearl Earring. *Herit. Sci.* **2020**, *8*, 1–16. [CrossRef]
11. Ricciardi, P.; Legrand, S.; Bertolotti, G.; Janssens, K. Macro X-ray fluorescence (MA-XRF) scanning of illuminated manuscript fragments: Potentialities and challenges. *Microchem. J.* **2016**, *124*, 785–791. [CrossRef]
12. Ricciardi, P.; Mazzinghi, A.; Legnaioli, S.; Ruberto, C.; Castelli, L. The Choir Books of San Giorgio Maggiore in Venice: Results of in Depth Non-Invasive Analyses. *Heritage* **2019**, *2*, 103. [CrossRef]
13. Clarke, M.L.; Gabrieli, F.; Rowberg, K.L.; Hare, A.; Ueda, J.; McCarthy, B.; Delaney, J.K. Imaging spectroscopies to characterize a 13th century Japanese handscroll, The Miraculous Interventions of Jizō Bosatsu. *Herit. Sci.* **2021**, *9*, 1–10. [CrossRef]
14. MacLennan, D.; Llewellyn, L.; Delaney, J.K.; Dooley, K.A.; Patterson, C.S.; Szafran, Y.; Trentelman, K. Visualizing and measuring gold leaf in fourteenth- and fifteenth-century Italian gold ground paintings using scanning macro X-ray fluorescence spectroscopy: A new tool for advancing art historical research. *Herit. Sci.* **2019**, *7*, 25. [CrossRef]

15. Legrand, S.; Van der Snickt, G.; Cagno, S.; Caen, J.; Janssens, K. MA-XRF imaging as a tool to characterize the 16th century heraldic stained-glass panels in Ghent Saint Bavo Cathedral. *J. Cult. Herit.* **2019**, *40*, 163–168. [CrossRef]
16. Lins, S.; Manso, M.; Lins, P.; Brunetti, A.; Sodo, A.; Gigante, G.; Fabbri, A.; Branchini, P.; Tortora, L.; Ridolfi, S. Modular MA-XRF Scanner Development in the Multi-Analytical Characterisation of a 17th Century *Azulejo* from Portugal. *Sensors* **2021**, *21*, 1913. [CrossRef]
17. Alfeld, M.; Baraldi, C.; Gamberini, M.C.; Walter, P. Investigation of the pigment use in the Tomb of the Reliefs and other tombs in the Etruscan Banditaccia Necropolis. *X-ray Spectrom.* **2019**, *48*, 262–273. [CrossRef]
18. Sottili, L.; Guidorzi, L.; Mazzinghi, A.; Ruberto, C.; Castelli, L.; Czelusniak, C.; Giuntini, L.; Massi, M.; Taccetti, F.; Nervo, M.; et al. The Importance of Being Versatile: INFN-CHNet MA-XRF Scanner on Furniture at the CCR "La Venaria Reale". *Appl. Sci.* **2021**, *11*, 1197. [CrossRef]
19. INFN-CHNet. Available online: http://chnet.infn.it/en/who-we-are-2/ (accessed on 29 April 2021).
20. Chiari, M.; Barone, S.; Bombini, A.; Calzolai, G.; Carraresi, L.; Castelli, L.; Czelusniak, C.; Fedi, M.E.; Gelli, N.; Giambi, F.; et al. LABEC, the INFN ion beam laboratory of nuclear techniques for environment and cultural heritage. *Eur. Phys. J. Plus* **2021**, *136*, 1–28. [CrossRef]
21. Giuntini, L.; Castelli, L.; Massi, M.; Fedi, M.; Czelusniak, C.; Gelli, N.; Liccioli, L.; Giambi, F.; Ruberto, C.; Mazzinghi, A.; et al. Detectors and Cultural Heritage: The INFN-CHNet Experience. *Appl. Sci.* **2021**, *11*, 3462. [CrossRef]
22. Campbell, L. Rogier van der Weyden and his Workshop. *Proc. Br. Acad.* **1994**, *84*, 1–24.
23. van Suchtelen, A. A painting from Florence, in Mauritshuis infocus. *Rogier van der Weyden Unvelied* **2018**, *2*, 14–15.
24. Kemperdick, S. Rogier van der Weyden's Workshop around 1440. In *Rogier van der Weyden in Context: Proceedings of Symposium XVII, Leuven, November 2009 (Underdrawing and Technology in Painting. Symposia)*; Campbell, L., Van Der Stock, J., Reynolds, C., Watteeuw, L., Eds.; Peeters Publishers: Paris, France; Leuven, Belgium; Walpole, MA, USA, 2012; pp. 57–77.
25. Billinge, R.; Campbell, L.; Dunkerton, J.; Foister, S.; Kirby, J.; Pilc, J.; Roy, A.; Spring, M.; White, R. The materials and technique of five paintings by Rogier van der Weyden and his Workshop. *Nat. Gallery Tech. Bull.* **1997**, *18*, 68–86.
26. Rogier van der Weyden Unveiled. Available online: https://www.mauritshuis.nl/en/discover/exhibitions/rogier-van-der-weyden-unveiled (accessed on 10 May 2021).
27. Taccetti, F.; Castelli, L.; Czelusniak, C.; Gelli, N.; Mazzinghi, A.; Palla, L.; Ruberto, C.; Censori, C.; Giudice, A.L.; Re, A.; et al. A multipurpose X-ray fluorescence scanner developed for in situ analysis. *Rend. Lince.* **2019**, *30*, 307–322. [CrossRef]
28. Ruberto, C.; Mazzinghi, A.; Massi, M.; Castelli, L.; Czelusniak, C.; Palla, L.; Gelli, N.; Betuzzi, M.; Impallaria, A.; Brancaccio, R.; et al. Imaging study of Raffaello's "La Muta" by a portable XRF spectrometer. *Microchem. J.* **2016**, *126*, 63–69. [CrossRef]
29. Mazzinghi, A.; Ruberto, C.; Castelli, L.; Ricciardi, P.; Czelusniak, C.; Giuntini, L.; Mandò, P.A.; Manetti, M.; Palla, L.; Taccetti, F. The importance of being little: MA-XRF on manuscripts on a Venetian island. *X-ray Spectrom.* **2021**, *50*, 272–278. [CrossRef]
30. Vadrucci, M.; Mazzinghi, A.; Sorrentino, B.; Falzone, S.; Gioia, C.; Gioia, P.; Loreti, E.M.; Chiari, M. Characterisation of ancient Roman wall-painting fragments using non-destructive IBA and MA-XRF techniques. *X-ray Spectrom.* **2020**, *49*, 668–678. [CrossRef]
31. Stols, M. Grounds. 1400–1900. In *Conservation of Easel Paintings*, 2nd ed.; Stoner, J.H., Rushfield, R., Eds.; Routledge Publisher: Oxfordshire, England, 2020; pp. 161–188, ISBN 9780367023799.
32. de Viguerie, L.; Glanville, H.; Ducouret, G.; Jacquemot, P.; Dang, P.; Walter, P. Re-interpretation of the Old Masters' practices through optical and rheological investigation: The presence of calcite. *Comptes Rendus Phys.* **2018**, *19*, 543–552. [CrossRef]
33. Spring, M. The Materials of Rogier van der Weyden and his contemporaries in context. In *Rogier van der Weyden in Context: Proceedings of Symposium XVII, Leuven, November 2009 (Underdrawing and Technology in Painting. Symposia)*; Campbell, L., van Der Stock, J., Reynolds, C., Watteeuw, L., Eds.; Peeters Publishers: Paris, France; Leuven, Belgium; Walpole, MA, USA, 2012; pp. 93–105.
34. Smieska, L.M.; Mullett, R.; Ferri, L.; Woll, A.R. Trace elements in natural azurite pigments found in illuminated manuscript leaves investigated by synchrotron x-ray fluorescence and diffraction mapping. *Appl. Phys. A* **2017**, *123*. [CrossRef]
35. Delaney, J.K.; Ricciardi, P.; Deming Glinsman, L.; Facini, M.; Thoury, M.; Palmer, M.; de la Rie, E.R. Use of imaging spectroscopy, fiber optic reflectance spectroscopy, and X-ray fluorescence to map and identify pigments in illuminated manuscripts. *Stud. Conserv.* **2014**, *59*, 91–101. [CrossRef]
36. Aru, M.; Burgio, L.; Rumsey, M.S. Mineral impurities in azurite pigments: Artistic or natural selection? *J. Raman Spectr.* **2014**, *45*, 1013–1018. [CrossRef]
37. Cennini, C. Chapter CXLVI, Come dèi fare vestiri di azzurro, d'oro, o di porpora. In *Il Libro dell'Arte, o Trattato della Pittura*; Milanesi, G., Milanesi, C., Eds.; Le Monnier: Firenze, Italy, 1859.
38. Keune, K.; Boon, J.J. Imaging Secondary Ion Mass Spectrometry of a paint cross section taken from an Early Netherlandish painting by Rogier van der Weyden. *Anal. Chem.* **2004**, *76*, 1374–1385. [CrossRef]
39. Ganio, M.; Pouyet, E.S.; Webb, S.M.; Schmidt Patterson, C.M.; Walton, M.S. From lapis lazuli to ultramarine blue: Investigating Cennino Cennini's recipe using sulfur K-edge XANES. *Pure Appl. Chem.* **2018**, *90*, 463–475. [CrossRef]
40. Gambardella, A.A.; Marine, C.; de Nolf, W.; Schnetz, K.; Erdmann, R.; van Elsas, R.; Gonzalez, V.; Wallert, A.; Iedema, P.D.; Eveno, M.; et al. Sulfur K-edge micro- and full-field XANES identify marker for preparation method of ultramarine pigment from lapis lazuli in historical paints. *Sci. Adv.* **2020**, *6*, eaay8782. [CrossRef]

41. Kirby, J.; Spring, M.; Higgitt, C. The Technology of Red Lake Pigment Manufacture: Study of the Dyestuff Substrate. *Natl. Gallery Tech. Bull.* **2005**, *26*, 71–87.
42. Lutzenberger, K.; Stege, H.; Tilenschi, C. A note on glass and silica in oil paintings from the 15th to the 17th century. *J. Cult. Herit.* **2010**, *11*, 365–372. [CrossRef]
43. Spring, M. New insights into the materials of fifteenth- and sixteenth-century Netherlandish paintings in the National Gallery, London. *Herit. Sci.* **2017**, *5*, 40. [CrossRef]
44. Mazzinghi, A.; Ruberto, C.; Castelli, L.; Czelusniak, C.; Taccetti, F.; Mandò, P.A. Indagini XRF a scansione sul Ritratto di papa Leone X con i cugini cardinali. In *Raffaello e il Ritorno del Papa Medici: Restauri e Scoperte sul Ritratto di Leone X con i due Cardinali*; Ciatti, M., Schmidt, E.D., Eds.; Edifir Edizioni Firenze: Firenze, Italy, 2020; Volume 54, pp. 219–226.
45. Wedepohl, K.H.; Simon, K. The chemical composition of medieval wood ash glass from Central Europe. *Geochemistry* **2010**, *70*, 89–97. [CrossRef]
46. Spring, M. Colourless Powdered Glass as an Additive in Fifteenth- and Sixteenth-Century European Paintings. *Natl. Gallery Tech. Bull.* **2012**, *33*, 4–26.
47. Gettens, R.J.; Feller, R.L.; Chase, W.T. Vermilion and Cinnabar. *Stud. Cons.* **1972**, *17*, 45–69.

Article

The Importance of Being Versatile: INFN-CHNet MA-XRF Scanner on Furniture at the CCR "La Venaria Reale"

Leandro Sottili [1,2], Laura Guidorzi [1,2], Anna Mazzinghi [3,4], Chiara Ruberto [3,4], Lisa Castelli [4], Caroline Czelusniak [4], Lorenzo Giuntini [3,4], Mirko Massi [4], Francesco Taccetti [4], Marco Nervo [5], Stefania De Blasi [5], Rodrigo Torres [6], Francesco Arneodo [6], Alessandro Re [1,2,*] and Alessandro Lo Giudice [1,2]

[1] Dipartimento di Fisica, Università degli Studi di Torino, Via Pietro Giuria 1, 10125 Torino, Italy; leandro.sottili@unito.it (L.S.); laura.guidorzi@unito.it (L.G.); alessandro.logiudice@unito.it (A.L.G.)

[2] Istituto Nazionale di Fisica Nucleare (INFN), Sezione di Torino, Via Pietro Giuria 1, 10125 Torino, Italy

[3] Dipartimento di Fisica e Astronomia, Università degli Studi di Firenze, Via Giovanni Sansone 1, Sesto Fiorentino, 50019 Firenze, Italy; anna.mazzinghi@unifi.it (A.M.); ruberto@fi.infn.it (C.R.); lorenzo.giuntini@unifi.it (L.G.)

[4] Istituto Nazionale di Fisica Nucleare (INFN), Sezione di Firenze, Via Giovanni Sansone 1, Sesto Fiorentino, 50019 Firenze, Italy; castelli@fi.infn.it (L.C.); czelusniak@fi.infn.it (C.C.); massi@fi.infn.it (M.M.); taccetti@fi.infn.it (F.T.)

[5] Centro Conservazione e Restauro "La Venaria Reale", Piazza della Repubblica, Venaria Reale, 10078 Torino, Italy; marco.nervo@centrorestaurovenaria.it (M.N.); stefania.deblasi@centrorestaurovenaria.it (S.D.B.)

[6] Division of Science, New York University Abu Dhabi, P.O. Box, Saadiyat Island, Abu Dhabi 129188, United Arab Emirates; rodrigo.torres@nyu.edu (R.T.); francesco.arneodo@nyu.edu (F.A.)

* Correspondence: alessandro.re@unito.it

Citation: Sottili, L.; Guidorzi, L.; Mazzinghi, A.; Ruberto, C.; Castelli, L.; Czelusniak, C.; Giuntini, L.; Massi, M.; Taccetti, F.; Nervo, M.; et al. The Importance of Being Versatile: INFN-CHNet MA-XRF Scanner on Furniture at the CCR "La Venaria Reale". *Appl. Sci.* **2021**, *11*, 1197. https://doi.org/10.3390/app11031197

Academic Editor: Anna Galli; Letizia Bonizzoni

Received: 31 December 2020
Accepted: 24 January 2021
Published: 28 January 2021

Publisher's Note: MDPI stays neutral with regard to jurisdictional claims in published maps and institutional affiliations.

Abstract: At present, the use of non-destructive, non-invasive X-ray-based techniques is well established in heritage science for the analysis and conservation of works of art. X-ray fluorescence (XRF) plays a fundamental role since it provides information on the elemental composition, contributing to the identification of the materials present on the superficial layers of an artwork. Whenever XRF is combined with the capability of scanning an area to provide the elemental distribution on a surface, the technique is referred to as macro X-ray fluorescence (MA-XRF). The heritage science field, in which the technique is extensively applied, presents a large variety of case studies. Typical examples are paintings, ceramics, metallic objects and manuscripts. This work presents an uncommon application of MA-XRF analysis to furniture. Measurements have been carried out with the MA-XRF scanner of the INFN-CHNet collaboration at the Centro di Conservazione e Restauro "La Venaria Reale", a leading conservation centre in the field. In particular, a chinoiserie lacquered cabinet of the 18th century and a desk by Pietro Piffetti (1701–1777) have been analysed with a focus on the characterisation of decorative layers and different materials (e.g., gilding in the former and ivory in the latter). The measurements have been carried out using a telemeter for non-planar surfaces, and with collimators of 0.8 mm and 0.4 mm diameter, depending on the spatial resolution needed. The combination of the small measuring head with the use of the telemeter and of a small collimator has guaranteed the ability to scan difficult-to-reach areas with high spatial resolution in a reasonable time (20×10 mm^2 with 0.2 mm step in less than 20 min).

Keywords: MA-XRF; conservation studies; furniture; Pietro Piffetti; chinoiserie lacquered cabinet

1. Introduction

X-ray fluorescence (XRF) is well established in the non-destructive, non-invasive analysis for the conservation, characterisation and prevention of works of art [1–5]. Whenever it is associated with the ability to scan an area, XRF provides the elemental composition related with the spatial distribution of the scanned area, and it is typically referred as macro X-ray fluorescence (MA-XRF) [1,4,6]. The MA-XRF technique is widely in use in the heritage science field, and a number of research groups have been developing customized instruments to improve the quality of data obtainable with this technique [6–9].

The MA-XRF technique may be provided by portable instruments or by large-scale facilities like synchrotrons [10–12]. Despite their higher performances in terms of output beam parameters (higher intensity, smaller spot size), synchrotron facilities have the limitation of needing the artifacts to be transported to the laboratory, which is not always feasible.

Among others, the Cultural Heritage Network of the National Institute of Nuclear Physics (INFN-CHNet) [13–15] is engaged in the development of instruments, methods and techniques for applications in the heritage science field [16–19]. With this intent, the INFN-CHNet collaboration has developed its own MA-XRF scanner. The main characteristics of this device are its compactness, light weight, user-friendly interface, and versatility as described in the next section. It has already been employed for a number of different applications, including paintings [20–22], illuminated manuscripts [23], coins [24], mosaics [25] and ceramics [24].

The INFN-CHNet MA-XRF scanner has recently been employed on furniture. This further application is particularly interesting since, at the present time, the literature on XRF analysis on furniture is still poor [26–28], and no cases of MA-XRF analysis on furniture have been reported.

In this paper, the application of the INFN-CHNet MA-XRF scanner on two different pieces of furniture of the 18th century is presented. One is a chinoiserie wooden cabinet placed at the *Castello e Parco di Masino*, property of the *Fondo Ambiente Italiano* (FAI) in Piedmont. The other is a writing desk *Scrivania con scansia* by Pietro Piffetti, placed in the *Palazzo Chiablese* in Turin, Piedmont.

Both works of art have been studied in a specific framework at the Centro di Conservazione e Restauro (CCR) "La Venaria Reale". The lacquered cabinet was part of the research project *"Un ponte tra l'Oriente e il Piemonte"*, a comparative study between lacquered oriental works of art of the 18th and 19th centuries, and their contemporary Western imitations. The main focus of the study was to determine the manufacturing techniques and materials employed in order to determine the elements required to date the manufactures and to distinguish between "original" oriental works of art and their Western replicas [29]. The study of the writing desk *Scrivania con scansia* by Pietro Piffetti from the *Palazzo Chiablese* has been realised within the research project on Pietro Piffetti carried out by the CCR [30,31]. This ongoing project is aimed at studying works of art by Pietro Piffetti and the Piedmontese cabinetmakers of the second half of the 18th century.

Thanks to the versatility of the INFN-CHNet MA-XRF scanner, it has been possible to study the decorative layers of the furniture, thus supporting their conservation processes and providing fundamental information to the research projects in which they are involved. The present work demonstrates the potentialities of the systematic use of a customized MA-XRF technique as a first approach to scientific studies on furniture.

2. Materials and Methods

The chinoiserie lacquered cabinet of the *Castello di Masino* (Figure 1b) is a typical example of furniture largely diffused throughout Europe in the second half of the 18th century. Historical research made it possible to date the furniture to 1780 and to attest to its production by a French workshop that was specialised in making furniture in imitation of oriental works assembling original Eastern lacquered panels. For this reason, it represents an important case study that permits the analysis of the aspects of both materials used in the Western imitations and those of the original oriental lacquers.

The writing desk *Scrivania con scansia* of *Palazzo Chiablese* (Figure 1a, $288 \times 155.5 \times 57.5$ cm^3) is a particularly significant work since it is the last documented masterpiece by Pietro Piffetti (Turin, 1701–1777), cabinetmaker of the Savoy court, dated to 1767. As part of the research on the Piedmontese cabinet-making, in particular on the furnishings by Pietro Piffetti, over the years, the CCR has been able to establish a database on the techniques and materials used, which today makes it possible to accurately document the evolution of the history of furniture technology, the basis of these great multi-material masterpieces of international furniture.

Figure 1. Pictures of the furniture and of the Cultural Heritage Network of the National Institute of Nuclear Physics (INFN-CHNet) macro X-ray fluorescence (MA-XRF) scanner during the measurements: (**a**) Writing desk *Scrivania con scansia* by Pietro Piffetti; (**b**) Chinoiserie wooden cabinet; (**c**) Scanner placed in front of the chinoiserie wooden cabinet; and (**d**) in front of the external side of the *scansia*.

The desk was created as a corner element of a boiserie inside the alcove hall of the *Palazzo Chiablese*. It is entirely inlaid and composed of veneers in violet rosewood and *bois de rose* and of polychrome engraved ivories and mother-of-pearl inlays. The cabinet-making furnishings of this period very rarely show inlays with polychrome engravings. Usually, they are made with black ink; hence, the possibility of mapping the nature of the inks and pigments present on this piece has provided an important element of comparison between the works that have been studied by the CCR in recent years [32,33].

The measurements have been carried out with the INFN-CHNet MA-XRF scanner at the CCR "La Venaria Reale". It is a compact ($60 \times 50 \times 50$ cm^3) and lightweight (around 10 kg) instrument completely assembled within the INFN-CHNet collaboration. Its main parts are the measuring head, composed by an X-ray tube (Moxtek© MAGNUM, 40 kV maximum voltage, 0.1 mA maximum anode current, Mo anode) with a collimator (changeable, typically between 400 μm and 2000 μm diameter), a silicon drift detector (Amptek© XR100 SDD, 50 mm^2 effective active surface, 140 eV FWHM at 5.9 keV), a telemeter (Keyence, model IA-100), three motor linear stages and a case containing all the electronics for acquisition and control. The motor stages (Physik Instrumente©, travel ranges 30 cm horizontally, x axis; 15 cm vertically, y axis; and 5 cm towards the specimen, z axis) holding the measuring head are screwed onto the carbon-fibre case containing the electronic components and the power supplies. The maximum operating voltage is 40 kV. Signals are collected with a multichannel analyser (model CAEN DT5780, also inside the carbon-fibre case), and the whole system is controlled by a laptop. The control–acquisition–analysis software is developed within the collaboration and allows both on-line and off-line analysis.

The output of the MA-XRF analysis is a file containing the scanning coordinates and, for each position, the spectrum acquired. As a result, the counts are recorded for each position. For each scanned area, or a part of it, a single element can be selected by its energy transition value and shown as an elemental 2D map. For each peak, the energy range is manually selected around the centroid according to its FWHM. The relative intensity of each element in a map is shown in greyscale, in which the maximum intensity is in white and the minimum in black.

Furthermore, multi-elemental maps can be created. In those maps, different elements are displayed in different colours (red, green, blue). This option permits the association of one or more pigments, by means of their chemical elements, to the visible features, allowing an immediate spatial distribution of the pigments (see, for example, Section 3.1).

A scan is carried out on the x axis, and a step size of 1 mm is typically set on the y axis, resulting in a pixel size of 1 mm^2. However, both the measuring speed and the pixel size can be adjusted depending on the need, as reported in the case of the writing desk. This ability, combined with the changeable dimension of the collimator, allows optimisation of the measuring time, and thus the maintenance of an adequate spatial resolution.

During measurements, the distance between the specimen and the measuring head is kept constant by means of the telemeter. The distance is set to the value of 6 mm, allowing the detector to point at the area irradiated by the source and to maximise the covered solid angle. Therefore, the detection efficiency is also improved for non-flat surfaces, especially for low-energy X-rays subjected to higher absorption by the air.

Moreover, for the detection of low-energy X-rays, helium flow may be conveyed between the target and the detector system from a nozzle installed close by the detector.

With the set-up described above, maps of elements with atomic numbers higher than Sodium ($Z > 11$) are efficiently provided by the instrument. A full review on the instrument can be found in [34].

3. Results and Discussion

3.1. Chinoiserie Lacquered Cabinet

For the investigation of this furniture, two areas, one for each panel, have been studied: one area corresponding to a flower (Figure 2), the other area to a flying bird (Figure 3).

For the two measurements, helium flow has been used. The scanning parameters are reported in Table 1.

Table 1. Scanning parameters of the areas of the chinoiserie lacquered cabinet.

Area	Dimension (mm^2)	Source Voltage (kV)	Anode Current (μA)	Scanning Speed (mm/s)	Collimator Diameter (μm)	Step Size (mm)
Flowers	145 × 105	28	30	3	800	1
Flying bird	175 × 135	28	30	3	800	1

In the first area, the query was related to the possible presence of arsenic-based compounds in the transparent yellow buds and flowers. Due to the overlap of the As Kα transition (10,544 keV) with the Pb Lα (10,552 keV), and of the As Kβ (11,726 keV) with the Hg Lβ (11,823 keV), the presence of arsenic can not be ascertained by a single elemental map. Therefore, maps of the major overlapping transitions among arsenic, mercury, and lead have been made and are reported in Figure 2d–f. As can be seen, even though the Hg

Lβ/As Kβ transition shows a high intensity in the red flower and the transparent yellow buds (Figure 2d), the Lα of Hg clearly shows its presence only in the red flower (Figure 2e), indicating that the signals in the area of the transparent yellow buds come from arsenic. The presence of the Pb Lβ transition has not been detected in the spectrum, confirming this result.

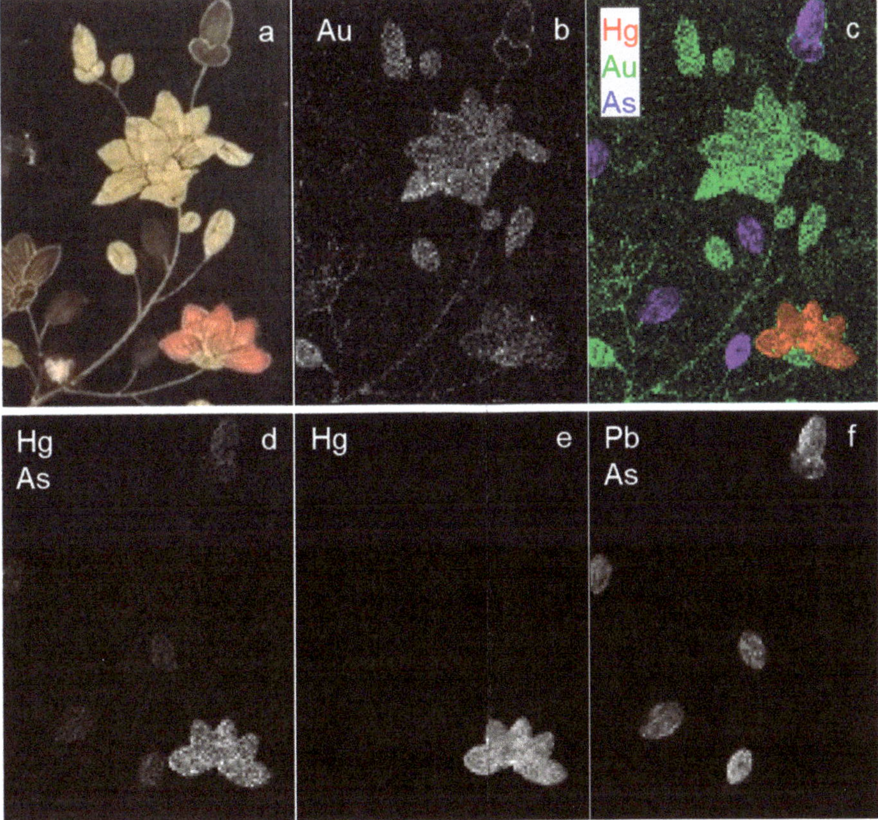

Figure 2. Elemental maps of the first area obtained by the INFN-CHNet MA-XRF scanner. (**a**) Visible. The maps presented are: (**b**) Au; (**c**) RGB map of Hg Lα (red), Au Lα (green) and As Kα (blue); (**d**) Hg Lβ/As Kβ transitions; (**e**) Hg Lα transition; (**f**) Pb Lα/As Kα transitions.

The presence of arsenic may be due to orpiment, but the use of realgar, pararealgar, or arsenic sulphide glass may not be excluded [35,36].

In order to clarify the composition of the different parts of the plant, an RGB map of Hg Lβ (red), Au Lα (green) and As Kα/Pb Lα (blue) transitions has been created and is reported in Figure 2c. As can be seen from Figure 2b, the bright yellow flowers, stems, outlines and highlights over red areas were made with gold. The red flower was likely realised with cinnabar-vermilion, of which the elemental composition is HgS [37].

The other area, even though it shows similarities with the previous, presents some essential differences in the decorative palette. The elemental maps of this area are shown in Figure 3. Compared with the area of the plant, the maps of mercury (Figure 3d) and gold (Figure 3g) attest to the probable use, respectively, of cinnabar-vermilion in the red tones, and of gold in the outline of the body and in one of the tails of the bird. The red tail is highlighted in gold as well.

As can be seen in the map of iron (Figure 3c), ochres-earths seem to have been used for the body of the bird. The eight black-blue feathers are related with a higher intensity of arsenic (Figure 3e), silicon (Figure 3h) and calcium (Figure 3b) compared with the rest of the area. In this case, with this technique it has not been possible to formulate a conclusive hypothesis on the decorative palette used by comparing the information with the present literature [38].

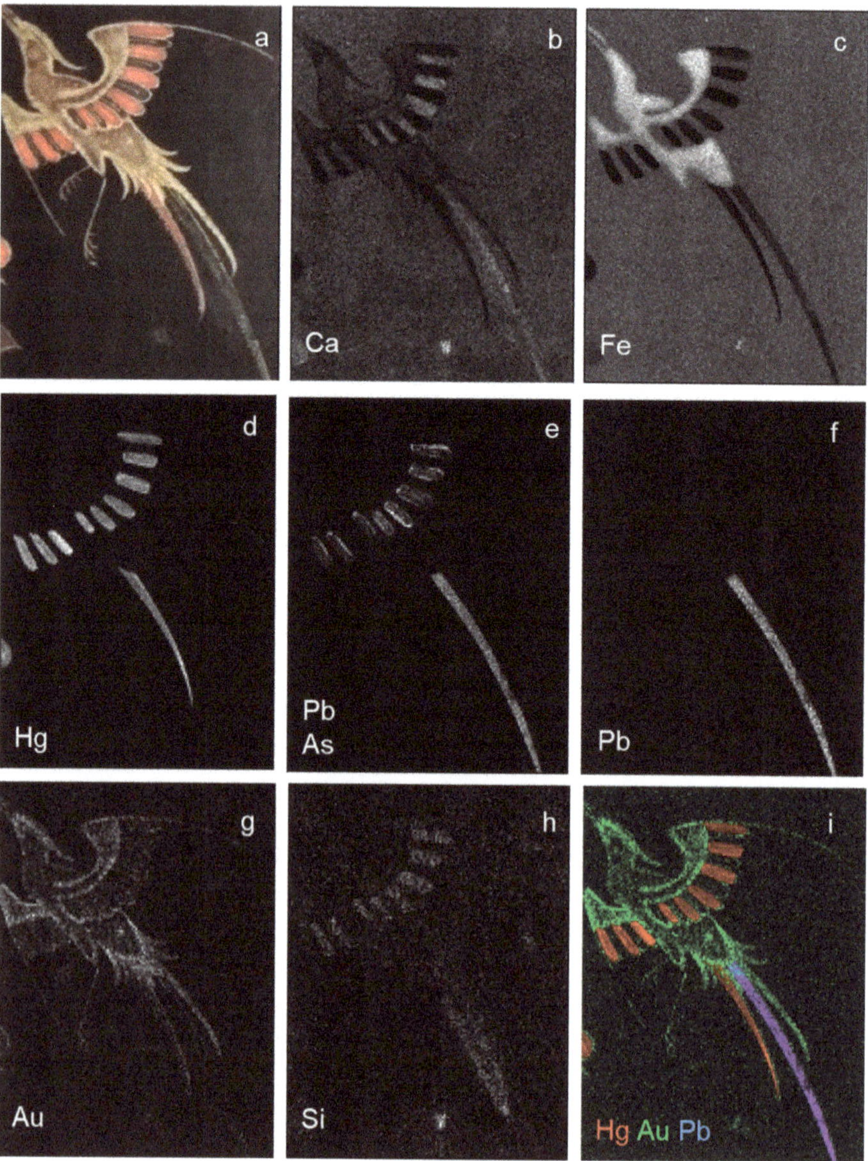

Figure 3. Elemental maps of the second area obtained by the INFN-CHNet MA-XRF scanner. (**a**) Visible. The maps presented are: (**b**) Ca Kα transition; (**c**) Fe Kα transition; (**d**) Hg Lα transition; (**e**) Pb Lα/As Kα transitions; (**f**) Pb Lβ transition; (**g**) Au Lα transition; (**h**) Si Kα transition; (**i**) RGB map of Hg (red), Au (green) and Pb (blue).

A different result has been achieved for the longest black tail, in which the presence of high Pb Lβ transition values reveals the use of Pb (Figure 3f). The literature [39,40] suggests that the presence of lead in black compounds may indicate the use of galena (PbS). The presence of sulphur can only be hypothesised since its transition (2.31 keV) overlaps with the transitions of mercury (M series) and lead (M series). However, with only XRF analysis, the use of another material like plattnerite (PbO_2) cannot be excluded. In addition, as can be seen from the map of silicon (Figure 3h), this element is present in the area of the tails, which may indicate a possible fourth tail with a similar composition to black-blue feathers. Finally, regarding the white small area below the red tail, the presence of a retouch is clearly visible in the maps of silicon (Figure 3h) and calcium (Figure 3b).

3.2. Writing Desk by Pietro Piffetti

The *Scrivania con scansia* by Pietro Piffetti has been analysed to characterise the decorating layers up on the marquetry. Three different areas were selected for this purpose: two areas on a drawer placed inside the writing desk, and one on a side, as shown in Figure 4. The choice was based on the hypotheses of different materials used and two possible conservation states inside and outside the writing desk.

Figure 4. Selected areas for the measurements: (**a**) selected areas outlined in yellow and the two points in the black and red shades on the drawer; (**b**) particular of the external side. The scanned areas are outlined in yellow.

The scanning parameters are reported in Table 2. Because of the small size of the details of the decoration, a better spatial resolution compared to the lacquered cabinet was required. Since the scanner has an available set of different collimators, for this application the smallest one, with the diameter of 400 μm, has been used. Thanks to this feature, a higher spatial resolution has been achieved in the elemental maps.

Table 2. Scanning parameters of the areas analysed on the *Scrivania con scansia* by Pietro Piffetti.

Area	Dimension (mm²)	Source Voltage (kV)	Anode Current (μA)	Scanning Speed (mm/s)	Collimator Diameter (μm)	Step Size (mm)
Drawer area 1	30 × 13	38	50	1	400	0.2
Drawer area 2	20 × 10	28	50	1	400	0.2
External side	40 × 130	28	70	1	400	0.2

The elemental maps of the first area of the drawer are shown in Figure 5. From the maps of lead (Figure 5c,d) it can be seen that its presence is detected in correspondence of the white highlights of the flower, most likely due to the use of lead white [41].

The shading in the central part of the flower is related to the presence of iron (Figure 5g) and manganese (Figure 5h), which may lead to the hypothesis of the use of ochres-earths [39]. In the green leaves, high signals of copper (Figure 5f) and arsenic (Figure 5d,e) are clearly visible. Even though this can be explained by the presence of a copper-based pigment or dye mixed with an arsenic-based compound, the literature suggests that this combination is an unlikely possibility, whereas the green colour may be related to the use of emerald green or Scheele's green [39,42]. However, it is reported that the latter has only been used since 1814, excluding its use for this piece of furniture; therefore, it is plausible that emerald green was used for the green leaves.

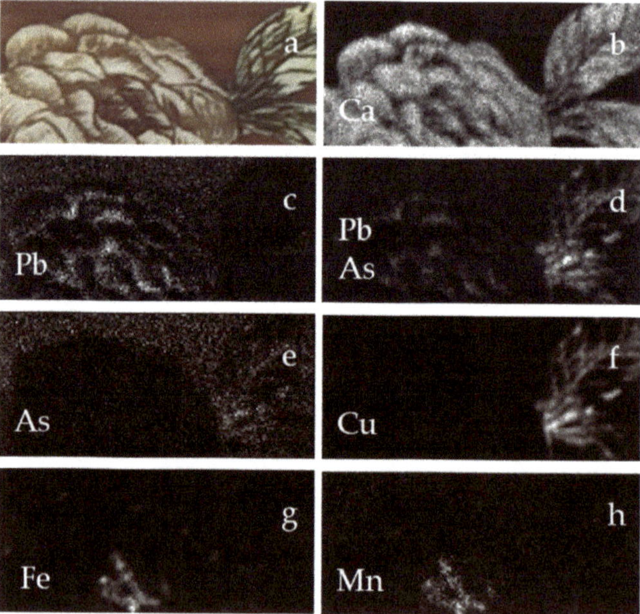

Figure 5. Elemental maps of the first area of the drawer obtained by the INFN-CHNet MA-XRF scanner. (**a**) Visible. The maps presented are: (**b**) Ca Kα transition; (**c**) Pb Lβ transition; (**d**) Pb Lα/As Kα transitions; (**e**) As Kβ transition; (**f**) Cu Kα transition; (**g**) Fe Kα transition; (**h**) Mn Kα transition.

The maps in the second area of the drawer, reported in Figure 6, confirm the same results of the first map, even though the black-blue flower on the top left shows a higher contribution of lead and iron. From the present literature, no single pigment is directly associated with the presence of those two elements [38,39]. Their presence can be explained by the use of Prussian blue combined with lead white, as reported in [43].

In the second area of the drawer shown in Figure 4, two points have been measured: one in the red tone and the other in the black tone. The measuring time was set to 120 s for each one to enhance the statistics of the spectra. As can be seen from Figure 7a, in the red tone the most intense peak is the calcium peak, probably related to the ivory beneath the decorative layer, whereas no other element with an atomic number above sodium was detected with relevant statistics. This result may indicate the use of organic dyes for this colour. The spectrum of the black-blue point (Figure 7b) shows a different elemental composition: the most prominent peaks are related to iron and lead (calcium is also present as a second peak in terms of intensity), confirming the result already discussed above.

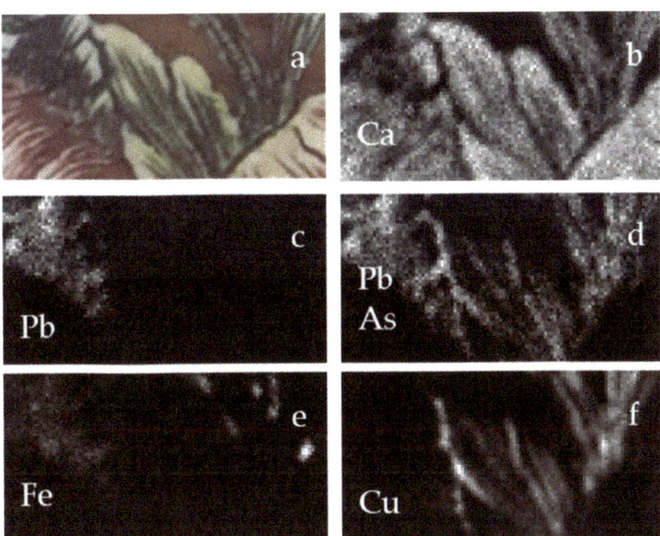

Figure 6. Elemental maps of the second area of the drawer. (**a**) Visible. The maps presented are: (**b**) Ca Kα transition; (**c**) Pb Lβ transition; (**d**) Pb Lα/As Kα transitions; (**e**) Fe Kα transition; (**f**) Cu Kα transition.

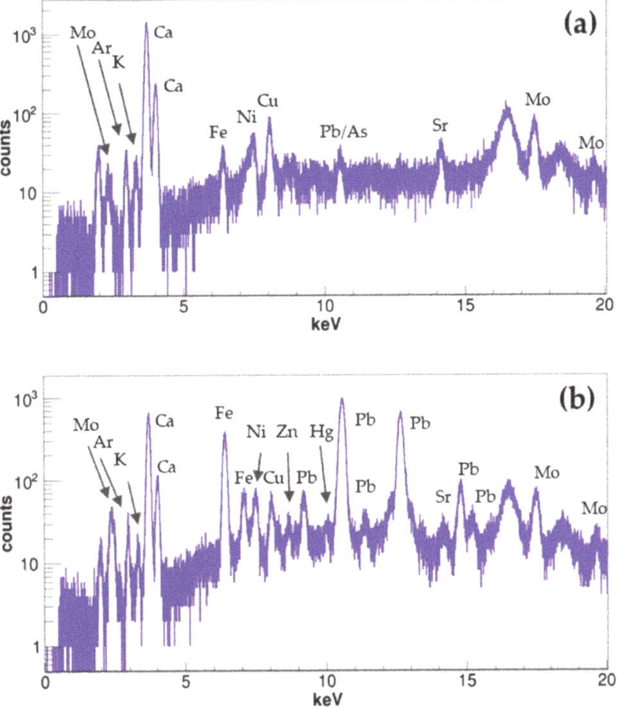

Figure 7. Spectra of the two points analysed with XRF: (**a**) red tone; (**b**) black tone. The presence of Ca probably due to ivory was detected in both.

On the external side, the two scans have been merged. As a result, elemental maps of the whole area indicated in Figure 4b were realised and are shown in Figure 8. In the area are green leaves, a dark blue-black stem and different flowers of red, black-blue and brown-orange tones. Some similarities with the drawer can be noted.

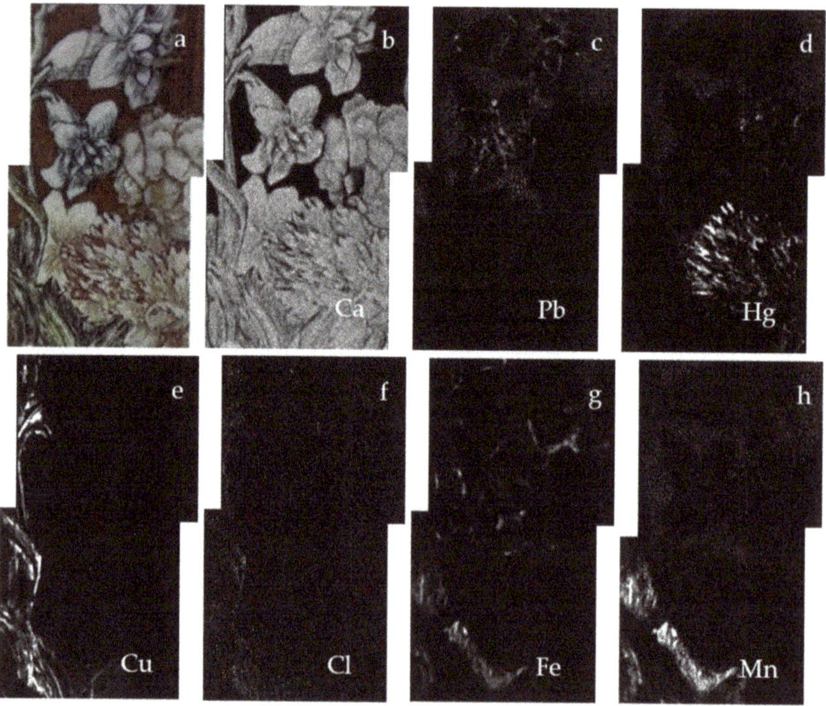

Figure 8. Elemental maps of the external area of the *Scrivania con scansia* obtained by the INFN-CHNet MA-XRF scanner. (**a**) Visible. The maps presented are: (**b**) Ca Kα transition; (**c**) Pb Lα transition; (**d**) Hg Lα transition; (**e**) Cu Kα transition; (**f**) Cl Kα transition; (**g**) Fe Kα transition; (**h**) Mn Kα transition.

In the shading of the red flower, mercury has been detected (Figure 8d), possibly due to the use of cinnabar-vermilion (HgS). Regarding the black-blue flowers, the same elements as in the black-blue flowers of the drawer—iron (Figure 8g) and lead (Figure 8c)—are present. Therefore, the decoration may have been realised with Prussian blue and lead white. In the blue-black stem area, the signals of iron and manganese are present in the elemental maps due to the probable use of iron-oxide-based compounds enriched with manganese oxides or other ochres-earths.

Regarding the dull-yellow flower, no map shows evidence of the presence of a characteristic element; therefore, the use of organic compounds can be hypothesised. Conversely, regarding the green parts in the drawer, a strong signal of copper (Figure 8e) associated with chlorine traces (Figure 8f) is present in the leaves on the side, whereas the presence of arsenic was not detected. This result led to the hypothesis of the use of different pigments in the green areas on the external side compared with the same colour in the drawer. Because the MA-XRF analysis does not allow the identification of compounds, it is not possible to determine which of the copper-based pigments has been used, or even if it was due to degradation effects as reported in [44].

A summary of the pigments hypothesised by the use of MA-XRF analysis is reported in Table 3.

Table 3. Summary of the hypothesised materials employed in the different areas.

Area Scanned	Colour	Elements Detected	Materials Hypothesised
Chinoiserie cabinet—flowers	red flower	Hg	cinnabar-vermilion
	bright yellow flowers	Au	gold
	transparent yellow buds	As	arsenic-based compound
Chinoiserie cabinet—flying bird	red tail and feathers	Hg	cinnabar-vermilion
	bright yellow outlines	Au	gold
	buff body	Fe	ochres-earths
	black-blue feathers	As, Si, Ca	?
	black tail	Pb	galena
Writing desk—drawer area 1	green leaves	As, Cu	emerald green
	dark rust tone of the flower	Fe, Mn	ochres-earths
	rust tone of the flower	-	organic compounds
	white highlights	Pb	lead white
Writing desk—drawer area 2	black-blue flower	Pb, Fe	Prussian blue and lead white
	green leaves	As, Cu	emerald green
Writing desk—external side	green leaves	Cu, Cl	copper-based compound
	black-blue flower	Pb, Fe	Prussian blue and lead white
	red flower	Hg	cinnabar-vermilion
	blue-black dark stem	Fe, Mn	ochres-earths
	dull yellow flower	-	organic compounds

4. Conclusions

The use of the MA-XRF technique on furniture has provided information on the elemental spatial distribution of the decorative layers. In particular, the INFN-CHNet MA-XRF scanner, due to its versatility in terms of hardware adaptability (different size of collimators, use of the telemeter) and in terms of software settings (scanning speed and step size during the measurements, data saving for an off-line analysis) has proved to be suitable for studies on this genre of works of art. A number of results have been achieved in identifying materials in the decorative layer. In both cases, the data on the polychromatic surfaces have provided information on the materials used: for instance, the use of arsenic-based compounds in the chinoiserie cabinet and the use of different pigments in the engraved ivories of the writing desk by Pietro Piffetti. However, the detection of only the elemental composition is a strong limitation for this kind of study. Therefore, the use of the technique is suggested as a first approach, while a multi-technical study is advisable for the identification of the compounds, as reported in [28]. For this reason, the INFN-CHNet group is involved in the development of instruments and techniques to support conservation processes, characterisation, and dating techniques in the heritage science field.

Author Contributions: Conceptualization, F.T., A.R. and A.L.G.; Data curation, L.S., L.G. (Laura Guidorzi), A.M. and C.R.; Formal analysis, L.S., L.G. (Laura Guidorzi), A.M. and C.R.; Funding acquisition, F.T. and A.L.G.; Investigation, L.G. (Laura Guidorzi), C.R., L.C., M.N., R.T. and A.R.; Methodology, L.G. (Laura Guidorzi), C.R., L.C., C.C., L.G. (Lorenzo Giuntini), M.M., F.T., M.N., R.T., F.A. and A.R.; Resources, M.N. and S.D.B.; Software, L.C., C.C. and F.T.; Supervision, A.R.; Visualization, L.S., L.G. (Laura Guidorzi), A.M. and C.R.; Writing—original draft, L.S.; Writing—review and editing, L.S., L.G. (Laura Guidorzi), A.M., A.R. and A.L.G. All authors have read and agreed to the published version of the manuscript.

Funding: The research was funded within the INFN-CHNet project. This project has received funding from the European Union's Horizon 2020 research and innovation programme under the Marie Skłodowska-Curie grant agreement No 754511 (PhD Technologies Driven Sciences: Technologies for Cultural Heritage—T4C).

Institutional Review Board Statement: Not applicable.

Informed Consent Statement: Not applicable.

Data Availability Statement: Not applicable.

Acknowledgments: The authors wish to warmly thank Marco Manetti for his suggestions, advice and invaluable technical support.

Conflicts of Interest: The authors declare no conflict of interest.

References

1. Alfed, M.; Broekaert, J.A.C. Mobile Depth Profiling and Sub-Surface Imaging Techniques for Historical Paintings—A review. *Spectrochiim. Acta Part B At. Spectrosc.* **2013**, *88*, 211–230. [CrossRef]
2. Angelici, D.; Borghi, A.; Chiarelli, F.; Cossio, R.; Gariani, G.; Lo Giudice, A.; Re, A.; Pratesi, G.; Vaggelli, G. μ-XRF analysis of trace elements in lapis lazuli-forming minerals for a provenance study. *Microsc. Microanal.* **2015**, *21*, 526–533. [CrossRef]
3. Corsi, J.; Lo Giudice, A.; Re, A.; Agostino, A.; Barello, F. Potentialities of X-ray fluorescence analysis in numismatics: The case study of pre-Roman coins from Cisalpine Gaul. *Archaeol. Anthropol. Sci.* **2018**, *10*, 431–438. [CrossRef]
4. Alfeld, M.; Siddons, D.P.; Janssens, K.; Dik, J.; Woll, A.; Kirkham, R.; van de Wetering, E. Visualizing the 17th century underpainting in Portrait of an Old Man by Rembrandt van Rijn using synchrotron-based scanning macro-XRF. *Appl. Phys. A* **2013**, *111*, 157–164. [CrossRef]
5. Alfeld, M.; Van der Snickt, G.; Vanmeert, F.; Janssens, K.; Dik, J.; Appel, K.; van der Loeff, L.; Chavannes, M.; Meedendorp, T.; Hendriks, E. Scanning XRF investigation of a Flower Still Life and its underlying composition from the collection of the Kröller–Müller Museum. *Appl. Phys. A* **2013**, *111*, 165–175. [CrossRef]
6. Romano, F.P.; Caliri, C.; Nicotra, P.; di Martino, S.; Pappalardo, L.; Rizzo, F.; Santos, H.C. Real-time elemental imaging of large dimension paintings with a novel mobile macro X-ray fluorescence (MA-XRF) scanning technique. *J. Anal. At. Spectrom.* **2017**, *32*, 773–781. [CrossRef]
7. Alfeld, M.; Pedroso, J.V.; van EikemaHommes, M.; Van der Snickt, G.; Tauber, G.; Blaas, J.; Haschke, M.; Erler, K.; Dik, J.; Janssens, K. A mobile instrument for in situ scanning macro-XRF investigation of historical paintings. *J. Anal. At. Spectrom.* **2013**, *28*, 760–767. [CrossRef]
8. Ravaud, E.; Pichon, L.; Laval, E.; Gonzalez, V.; Eveno, M.; Calligaro, T. Development of a versatile XRF scanner for the elemental imaging of paintworks. *Appl. Physics. A* **2015**, *122*. [CrossRef]
9. Alberti, R.; Frizzi, T.; Bombelli, L.; Gironda, M.; Aresi, N.; Rosi, F.; Miliani, C.; Tranquilli, G.; Talarico, F.; Cartechini, L. CRONO: A fast and reconfigurable macro X-ray fluorescence scanner for in-situ investigations of polychrome surfaces. *X-ray Spectrom.* **2017**, *46*. [CrossRef]
10. Bergmann, U.; Bertrand, L.; Edwards, N.P.; Manning, P.L.; Wogelius, R.A. *Chemical Mapping of Ancient Artifacts and Fossils with X-ray Spectroscopy*; Jaeschke, E., Khan, S., Schneider, J., Hastings, J., Eds.; Synchrotron Light Sources and Free-Electron Lasers; Springer: Cham, Switzerland, 2019. [CrossRef]
11. Debastiani, R.; Simon, R.; Batchelor, D.; Dellagustin, G.; Baumbach, T.; Fiederle, M. Synchrotron-based scanning macro-X-ray fluorescence applied to fragments of Roman mural paintings. *Microchem. J.* **2016**, *126*, 438–445. [CrossRef]
12. Bergmann, U.; Manning, P.L.; Wogelius, R.A. Chemical mapping of paleontological and archeological artifacts with synchrotron X-Rays. *Annu. Rev. Anal. Chem.* **2012**, *5*, 361–389. [CrossRef] [PubMed]
13. Fernandez, J.E.; Taccetti, F.; Kenny, J.M.; Amendola, R. Conclusive editorial on non-destructive techniques for cultural heritage. *Rend. Fis. Acc. Lincei* **2020**, *31*, 819–820. [CrossRef]
14. Castelli, L.; Felicetti, A.; Proietti, F. Heritage Science and Cultural Heritage: Standards and tools for establishing cross-domain data interoperability. *Int. J. Digit Libr.* **2019**. [CrossRef]
15. Sottili, L.; Guidorzi, L.; Mazzinghi, A.; Ruberto, C.; Castelli, L.; Czelusniak, C.; Giuntini, L.; Massi, M.; Taccetti, F.; Nervo, M.; et al. INFN-CHNet meets CCR La Venaria Reale: First results. In Proceedings of the 2020 IMEKO TC-4 International Conference on Metrology for Archaeology and Cultural Heritage 2020, Trento, Italy, 22–24October 2020; pp. 507–511.
16. Re, A.; Zangirolami, M.; Angelici, D.; Borghi, A.; Costa, E.; Giustetto, R.; Gallo, L.M.; Castelli, L.; Mazzinghi, A.; Ruberto, C.; et al. Towards a portable X-Ray Luminescence instrument for applications in the Cultural Heritage field. *Eur. Phys. J. Plus* **2018**, *133*, 362. [CrossRef]
17. Czelusniak, C.; Palla, L.; Massi, M.; Carraresi, L.; Giuntini, L.; Re, A.; Lo Giudice, A.; Pratesi, G.; Mazzinghi, A.; Ruberto, C.; et al. Preliminary results on time-resolved ion beam induced luminescence applied to the provenance study of lapis lazuli. *Nucl. Instrum. Methods Phys. Res. B* **2016**, *371*, 336–339. [CrossRef]
18. Palla, L.; Czelusniak, C.; Taccetti, F.; Carraresi, L.; Castelli, L.; Fedi, M.E.; Giuntini, L.; Maurenzig, P.R.; Sottili, L.; Taccetti, N. Accurate on line measurements of low fluences of charged particles. *Eur. Phys. J. Plus* **2015**, *130*. [CrossRef]
19. Lo Giudice, A.; Corsi, J.; Cotto, G.; Mila, G.; Re, A.; Ricci, C.; Sacchi, R.; Visca, L.; Zamprotta, L.; Pastrone, N. A new digital radiography system for paintings on canvas and on wooden panels of large dimensions. In *Proceedings of the 2017 IEEE International Instrumentation and Measurement Technology Conference (I2MTC 2017) Proceedings*; IEEE: Piscataway, NJ, USA, 2017; pp. 1834–1839, FP17IMT-ART; ISBN 9781509035960.
20. Ruberto, C.; Mazzinghi, A.; Massi, M.; Castelli, L.; Czelusniak, C.; Palla, L.; Gelli, N.; Bettuzzi, M.; Impallaria, A.; Brancaccio, R.; et al. Imaging study of Raffaello's "La Muta" by a portable XRF spectrometer. *Microchem. J.* **2016**, *126*, 63–69. [CrossRef]

21. Vadrucci, M.; Mazzinghi, A.; Sorrentino, B.; Falzone, S.; Gioia, C.; Gioia, P.; Loreti, E.M.; Chiari, M. Characterisation of ancient Roman wall-painting fragments using non-destructive IBA and MA-XRF techniques. *X-Ray Spectrom.* **2020**, *49*, 668–678. [CrossRef]

22. Dal Fovo, A.; Mazzinghi, A.; Omarini, S.; Pampaloni, E.; Ruberto, C.; Striova, J.; Fontana, R. Non-invasive mapping methods for pigments analysis of Roman mural paintings. *J. Cult. Herit.* **2020**, *43*, 311–318. [CrossRef]

23. Mazzinghi, A.; Ruberto, C.; Castelli, L.; Ricciardi, P.; Czelusniak, C.; Giuntini, L.; Mandò, P.A.; Manetti, M.; Palla, L.; Taccetti, F. The importance of being little: MA-XRF on manuscripts on a Venetian island. *X-ray Spectrom* **2020**, 1–7, in press. [CrossRef]

24. Lazic, V.; Vadrucci, M.; Fantoni, F.; Chiari, M.; Mazzinghi, A.; Gorghinian, A. Applications of laser-induced breakdown spectroscopy for cultural heritage: A comparison with X-ray Fluorescence and Particle Induced X-ray Emission techniques. *Spectrochim. Acta Part B At. Spectrosc.* **2018**, *149*, 1–14. [CrossRef]

25. Mazzinghi, A. Sviluppo di Strumentazione XRF a Scansione per Applicazioni ai Beni Culturali. Ph.D. Thesis, University of Florence, Firenze, Italy, 2016.

26. Andersson, E.; Cattersel, V. A Dutch Seventeenth-Century European Lacquer Cabinet. Material-Technical Analysis to Gain Insight into the Deteriorated Surface. In *Material Imitation and Imitation Materials in Furniture and Conservation*; Vasques Dias, M., Ed.; Stichting Ebenist: Amsterdam, The Netherlands, 2017; pp. 190–206.

27. Salvemini, F.; Grazzi, F.; Agostino, A.; Iannaccone, R.; Civita, F.; Hertmann, S.; Lehmann, E.; Zoppi, M. Non-invasive characterization through X-ray fluorescence and neutron radiography of an ancient Japanese lacquer. *Archaeol. Anthropol. Sci.* **2013**, *5*, 197–204. [CrossRef]

28. Felix, V.S.; Mello, U.L.; Pereira, M.O.; Oliveira, A.L.; Ferreira, D.S.; Carvalho, C.S.; Silva, F.L.; Pimenta, A.R.; Diniz, M.G.; Freitas, R.P. Analysis of a European cupboard by XRF, Raman and FT-IR. *Radiat. Phys. Chem.* **2018**, *151*, 198–204. [CrossRef]

29. Tagliante, S. Problematiche Conservative e Restauro di uno Stipo Settecentesco con Decorazioni in Lacca Orientale e "Alla China". Master's Thesis, University of Turin, Torino, Italy, 2018.

30. Spantigati, C.; De Blasi, S. *Il Restauro degli Arredi Lignei. L'ebanisteria Piemontese. Studi e Ricerche*; Nardini Editore: Firenze, Italy, 2011.

31. Re, A.; Albertin, F.; Avataneo, C.; Brancaccio, R.; Corsi, J.; Cotto, G.; De Blasi, S.; Dughera, G.; Durisi, E.; Ferrarese, W.; et al. X-ray tomography of large wooden artworks: The case study of "Doppio corpo" by Pietro Piffetti. *Herit. Sci.* **2014**, *2*. [CrossRef]

32. De Blasi, S.; Nervo, M.; Ravera, M.; Spantigati, C. Structural characters of Piedmontese eighteenth-century cabinetmaking: Historical documents, restorations and new technologies, in restoring joints, conserving structures. In Proceedings of the Tenth International Symposium on Wood and Furniture Conservation, Amsterdam, The Netherlands, 8–9 October 2010; pp. 98–107.

33. Luciani, P.; De Blasi, S.; Nervo, M.; Piccirillo, A. *Il Restauro del Mobile di Ebanisteria Piemontese del Settecento: Le Opere di Pietro Piffetti e Luigi Prinotto. Alcuni Casi Studio, Conservació-Restauració del Moble i la Fusta. L'experiència dels Experts*; Costa Galobart, N., Ed.; Associació per a l'Estudi del Moble: Ajuntament de Barcelona, Institut de Cultura, Museu del Disseny de Barcelona: Barcelona, Spain, 2020; pp. 85–98.

34. Taccetti, F.; Castelli, L.; Czelusniak, C.; Gelli, N.; Mazzinghi, A.; Palla, L.; Ruberto, C.; Censori, C.; Lo Giudice, A.; Re, A.; et al. A multipurpose X-ray fluorescence scanner developed for in situ analysis. *Rend. Fis. Acc. Lincei* **2019**, *30*, 307–322. [CrossRef]

35. Fitzhugh, E. *West, Orpiment and Realgar in Artists Pigments*; West Fitzhugh, E., Ed.; National Gallery of Art: London, UK, 1997; Volume 3, pp. 45–81.

36. Vermeulen, M.; Sanyova, J.; Janssens, K. Identification of artificial orpiment in the interior decorations of the Japanese tower in Laeken. *Bruss. Belg. Herit. Sci.* **2015**, *3*, 9. [CrossRef]

37. Gettens, R.J.; Robert, L.F.; Chase, W.T. Vermilion and Cinnabar. *Stud. Conserv.* **1972**, *17*, 45–69. [CrossRef]

38. Marika, S.; Grout, R.; White, R. 'Black Earths': A Study of Unusual Black and Dark Grey Pigments Used by Artists in the Sixteenth Century. *Natl. Gallery Tech. Bull.* **2003**, *24*, 96–114.

39. Seccaroni, C.; Moioli, P. *Fluorescenza X-Prontuario per l' analisi XRF Portatile Applicata a Superfici Policrome*; Nardini Editore: Firenze, Italy, 2002; pp. 60–89.

40. Siddall, R. Mineral Pigments in Archaeology: Their Analysis and the Range of Available Materials. *Minerals* **2018**, *8*, 201. [CrossRef]

41. Eastaugh, N. *Pigment Compendium: A Dictionary and Optical Microscopy of Historical Pigments*; Butterworth-Heinemann: Amsterdam, The Netherlands, 2008; pp. 239–241.

42. Kriznar, A.; Muñoz, M.; Paz, F.; Respaldiza, M.; Vega, M. Non-destructive XRF analysis of pigments in a 15th century panel painting. In Proceedings of the 9th International Conference on NDT of Art, Jerusalem, Israel, 25–30 May 2008.

43. Kirby, J.; Saunders, D. Fading and colour change of Prussian blue: Methods of manufacture and the influence of extenders. *Natl. Gallery Tech. Bull.* **2004**, *25*, 73–99.

44. Vermeulen, M.; Sanyova, J.; Janssens, K.; Nuyts, G.; De Meyer, S.; De Wael, K. The Darkening of Copper- or Lead-Based Pigments Explained by a Structural Modification of Natural Orpiment: A Spectroscopic and Electrochemical Study. *J. Anal. At. Spectrom.* **2017**, *32*, 1331–1341. [CrossRef]

Article

More than XRF Mapping: STEAM (Statistically Tailored Elemental Angle Mapper) a Pioneering Analysis Protocol for Pigment Studies [†]

Jacopo Orsilli [1], Anna Galli [1,2,*], Letizia Bonizzoni [3,*] and Michele Caccia [1]

1   Dipartimento di Scienza dei Materiali, Università degli Studi Milano-Bicocca, Via R. Cozzi 55, 20125 Milano, Italy; j.orsilli@campus.unimib.it (J.O.); michele.caccia@unimib.it (M.C.)
2   CNR-IBFM, Via Fratelli Cervi 93, 20090 Segrate (MI), Italy
3   Dipartimento di Fisica Aldo Pontremoli, Università degli Studi di Milano, Via Celoria 16, 20133 Milano, Italy
*   Correspondence: anna.galli@unimib.it (A.G.); letizia.bonizzoni@unimi.it (L.B.)
†   In loving memory of Professor Mario Milazzo.

Academic Editor: Radian Popescu

Received: 13 January 2021
Accepted: 2 February 2021
Published: 5 February 2021

Abstract: Among the possible variants of X-Ray Fluorescence (XRF), applications exploiting scanning Macro-XRF (MA-XRF) are lately widespread as they allow the visualization of the element distribution maintaining a non-destructive approach. The surface is scanned with a focused or collimated X-ray beam of millimeters or less: analyzing the emitted fluorescence radiation, also elements present below the surface contribute to the elemental distribution image obtained, due to the penetrative nature of X-rays. The importance of this method in the investigation of historical paintings is so obvious—as the elemental distribution obtained can reveal hidden sub-surface layers, including changes made by the artist, or restorations, without any damage to the object—that recently specific international conferences have been held. The present paper summarizes the advantages and limitations of using MA-XRF considering it as an imaging technique, in synergy with other hyperspectral methods, or combining it with spot investigations. The most recent applications in the cultural Heritage field are taken into account, demonstrating how obtained 2D-XRF maps can be of great help in the diagnostic applied on Cultural Heritage materials. Moreover, a pioneering analysis protocol based on the Spectral Angle Mapper (SAM) algorithm is presented, unifying the MA-XRF standard approach with punctual XRF, exploiting information from the mapped area as a database to extend the comprehension to data outside the scanned region, and working independently from the acquisition set-up. Experimental application on some reference pigment layers and a painting by Giotto are presented as validation of the proposed method.

Keywords: MA-XRF; SAM; STEAM; pigments; elemental mapping; painting stratigraphy; Giotto

1. Introduction

Until about ten years ago, talking about X-Ray Fluorescence (XRF) imaging meant talking about Synchrotron Radiation of advanced generation for tomography and holography [1]. Nowadays, XRF imaging in Cultural Heritage (CH) field immediately recalls XRF mapping (or Macro-XRF, MA-XRF), which is the recording of XRF spectra together with their spatial coordinates, so to allow the reconstruction of distribution maps for chemical elements in the sample. This can be obtained, in principle, by either scanning the sample or using a pinhole and a 2D detector, the so-called full-field XRF (FF-XRF) [2–4].

MA-XRF mobile instruments scan samples, as painted artworks, with an X-ray beam whose dimensions can vary from one to several hundred micrometers in diameter. The detected characteristic X-ray fluorescence emission, stimulated by ingoing radiation, allows creating the images of the elemental distribution. Technical details are beyond the purpose of this paper, aiming to discuss the consequences in CH field and the possibilities linked to a good post-processing of data. Indeed, the physical background of MA-XRF does not differ from that of XRF, but the possibility of mapping elemental distribution on samples gives this technique the dignity of a new analytical method and makes it one of the most useful tools to support restoration and conservation interventions. One of the characteristics which makes MA-XRF a routine technique not only on painting or CH materials but also in many other fields (such as geology, biology, forensic [5–10]), is the multispectral capabilities and the possibility of visualizing the distribution of chemical elements at and below the surface of paintings, according to the self-absorption limits for X-rays. Indeed, maps are a more intuitive tool that can be easily interpreted by outsiders, such as restorers, art-historians, curators, and conservators.

The first experiments in 2007 were performed with synchrotron radiation [11]. Ten years until then, MA-XRF regularly applied in museums, archeological sites, conservation studios, and the first workshop on MA-XRF, Macro X-ray Fluorescence Scanning in Conservation, Art, and Archaeology was held in Trieste [12].

Nowadays, portable MA-XRF (p-MA-XRF) has moved over, with the possibility of taking into account also the shape of the scanned surface [13,14] and combining its results with all the other well-established imaging techniques [15–21]. Recent instrumentation organizes data in 3D spectral image cubes (one of the axes is the spectral one, the other two axes describe the position) [22]. Moreover, and this is the focus of our paper, data analysis has been more and more developed and optimized [23–25] until the interpretation using neural networks [26].

1.1. State of the Art Instruments and Methods

Cultural Heritage artifacts are known to have a high variability: they have different sizes, different shapes, and are made of very different materials. So, as adaptation is the most important characteristic of a living being, the same can be said for modern MA-XRF spectrometers. That is why analytical techniques have shown a high development in their portable versions, without neglecting the quality of the results. Indeed, modern spectrometers are required to analyze all kinds of samples with higher sensitivity, both spatial and analytical, adapting to the characteristic of each single artifact, sometimes also coupling multiple techniques.

The development of MA-XRF scanners has focused mostly on reducing the dwell time and on increasing the spatial resolution, as these two parameters are indeed those that define the measurement time of a surface. The dwell time represents the time required for the analysis of a pixel, while the spatial resolution represents the size of the pixel so that the total measurement time can be described as:

$$T = (d_t + m_t) N \tag{1}$$

where d_t is the dwell time, m_t is the time needed to move to the next pixel and N is the number of pixels. The higher is the spatial resolution, the higher is the number of pixels needed to cover the same area; the lower is the dwell time, the lower is the time of analysis. To exemplify the high improvements made in the last thirty years, we just need to note that one of the first developed portable instruments had a dwell time of 30 s (plus 25–30 s to move to the next sample point) and a best spatial resolution of 1 mm so that hours were needed to analyze a small portion of a painting [27]. For modern instruments, dwell time is reduced by 1–4 orders of magnitudes, and an improvement of 2 orders of magnitude in spatial resolution is obtained, so that the same areas can be analyzed with a higher resolution in less time. To improve the spatial resolution, capillaries and capillary lens substitute the collimators, ensuring a higher flux. The signal intensity has also been increased using suitable detectors, especially Silicon Drift Detectors (SDDs) [24,28–30]. Optimization of geometrical conditions have also been permitted to reduce the dwell time: minimizing the distance of the detector from the sample, the absorption of air at low energies is reduced, and at the same time, the acceptance solid angle is increased [31]. Furthermore, detectors with a larger detection area (or more than one detector in combination) highly increase the detected signal, speeding the measurement time up.

Moreover, the size and the portability of these system allows to analyze samples with different dimensions, from manuscripts to large paintings [32–36], not only in museums or archives but also in archaeological sites [37], extending in this way the advantages of this kind of analysis to all the possible objects of the CH world. Besides, the higher spatial resolution has opened new possibilities for this technique, which allows not only to analyze the composition of the artist palette, or the presence of underdrawings, but that is also capable of recovering degraded and illegible daguerreotypes [38].

1.2. State of the Art. Data Handling and Synergic Applications

The data collected through an MA-XRF analysis are a $x \times y \times N$ matrices, where $x \times y$ are the pixels (sample points) collected and N are the channels of the ED-XRF (Energy Dispersive-XRF) spectra. The analysis of each spectrum is not different from the analysis of the same data collected with other XRF methods, i.e., punctual XRF, thus the data treatment of each spectrum can follow the same protocol both for a qualitative and for a quantitative analysis. Yet, we have to face two main problems: the amount of data to be analyzed, and the inhomogeneity of the data itself, that is the variation of the data caused by other parameters than sample composition, as variation in the spectrometer–sample distance, shadowing (of the source and the detectors) due to the sample shape, and changes in the roughness of the surface [39–41]. These sources of error must be taken into account to perform a proper data reduction without compromising their correct evaluation.

Some instruments can automatically maintain a constant distance between the spectrometer and the sample: this is particularly important when dealing with large paintings, usually warped, or when analyzing 3D objects, like jewels or other kinds of artifacts [27–29]. Indeed, if the distance is allowed to vary, a post-processing of the spectra is required, especially if we need to analyze light elements, highly affected by air absorption. A way to correct the data is proposed by Alfeld et al. [13], exploiting the argon signal, due only to the interaction of X-rays with air and, as a first approximation, proportional to the sample–detector distance. Even if this method has been applied to stitch multiple MA-XRF images of the same paintings, it may also be used to correct small variations in the acquisition distance.

The data correction for the shadowing effect is more complex: when the analyzed spot is partially hidden from the detector due to the morphology of the artifact, the artifact itself causes a reduction of the solid angle of the detector/source, lowering the signal. Some spectrometers mount two detectors in opposite directions [24] that partially mitigate this effect.

If some of these sources of errors can neither be corrected in the post-processing of the spectra, they must be taken into account during the data analysis, with particular attention to the morphology of the sample.

As already said, the amount of data collected in a MA-XRF scanning is a cubic matrix with hundreds of spectra; it is obvious that an operator cannot analyze all the spectra. A data reduction process is thus necessary and can be performed both as a channel data reduction (reducing the matrix along the N axis) or as a sample data reduction (reducing the matrix along the $x \times y$ axes).

The channel data reduction is probably the simplest to perform, for instance, converting the spectra into their fitted data, retrieving only the net area of the peaks. This process can be performed through the analysis of the sum spectrum of all the $x \times y$ sample points, to evaluate the elements present in the whole scanning to be fitted in the following spectral analysis. The matrix thus obtained can then be treated by simple statistical methods, such as Principal Component Analysis (PCA), probably one of the most used techniques for data reduction, especially thanks to the low data loss that it provides. The N channels can then be represented in a k-dimensional space (where k is lower than N) and grouped into clusters.

Different approaches relying on clustering methods can be used for sample data reduction. Most of these methods aim to cluster similar spectra in groups: each group can then be represented by an endmember, which summarizes the characteristics (the chemical composition) of the whole group.

A way to cluster XRF spectra is the use of SAM (Spectra Angle Mapper): in this method, a spectrum is considered as a point in a space of N dimensions, where N is the number of the channels. The difference between two spectra can be evaluated as the angle that divides the vectors heading to the two N-dimensional points:

$$\alpha_{(i,j)_k} = \arccos\left(\frac{\sum_{l=1}^{N} \rho_{kl}\rho'_{(i,j)_l}}{L_{\rho k}L_{\rho'(i,j)}}\right) \tag{2}$$

where ρ_l and ρ'_l are the values of the channels of the two spectra to confront. This technique has been widely used for the analysis of multispectral data, but also for XRF data [42,43]. For each group, an endmember can be found, which exemplifies the chemical composition of all the sample points of the same group.

Another way to obtain the same results is the use of Artificial Neural Networks (ANNs) [44,45]. Usually, ANNs require training to calculate the weights associated with each neuron, and then the weights can be employed to analyze the dataset. In the work of Kogou et al. [26], the authors use a Self-Organizing Map (SOM) method that skips the training process using the same dataset as a training set, allowing a fully automated clustering.

As an example of the flexibility and of the enormous potential of XRF mapping, we are introducing and discussing a pioneering analysis protocol called STEAM, Statistically Tailored Elemental Angle Mapper. Based on the SAM algorithm and on the Global Spectrum (GS) analysis introduced by Galli and co-workers [46], STEAM first accomplishes the tasks of the MA-XRF typical approach (i.e., it maps the elemental distribution [10,14]) and then, in synergy with punctual XRF, it exploits the information gathered from the mapped area(s) as a database to extend the comprehension to data outside the scanned region(s). The protocol works independently from the acquisition set-up and, once defined the main features of the database, it sheds light on uncharacterized data, disregarding their origin and their relationships with the scanned region(s).

2. Materials and Methods

2.1. Instrumentation

The experimental data are X-ray fluorescence spectra, collected with two different spectrometers, ELIO by XGLab srl, Bruker Nano Analytics Division, and ARTAX 200 by Bruker. Even if the operative range of both the instruments is wider than 2.5–14.5 keV, the analyzed data were limited to this interval to exclude from the spectra the emission lines of both the X-ray excitation tubes; in fact, spectrometers employ anodes in Rhodium (ELIO) or Molybdenum (ARTAX 200), two elements that present K- and L-series at energies higher than 15 keV and lower than 2.5 keV.

2.1.1. ELIO Spectrometer by XGLab Srl

This spectrometer is based on 25 mm^2 area, Peltier-cooled, Silicon Drift Detector characterized by an energy resolution of 130 eV on the Mn-k$_\alpha$ line at 0.5 μs shaping time, a peak-to-background ratio of the order of 15,000, and by silicon thickness of 500 μm. The excitation source is a low power (4 W), 50 kV, transmission X-ray tube with Rh anode, coupled to a 1 mm collimator. The source exploits interchangeable filters and provides a 1.2 mm spatial resolution on the sample at the focusing position. The design of the source and detection system and the large solid angle of X-ray collection are optimized to obtain a high counting-rate capability even if a low power X-ray tube is used. The compact detection head is completed by an integrated microscope camera (magnification $\times$10, field of view about 8 $\times$ 6 mm), an external video-camera, focal and axial lasers for fine positioning, and LEDs light for sample illumination on the analysis region with dimming adjustment. The system can be equipped with an XY translation stage (with 100 mm total travel and 100 μm step size), for 2D mapping. For a typical XRF map (50 × 50 mm of scanned area, 1 mm step, and 1 s acquisition per step) the total measurement time is less than 1 h. Since the excitation scheme does not require poly-capillary optics, the spectrometer is highly sensitive in the full range of 1–40 keV. A tripod with 3D orientation makes the instrument portable and compatible with almost any measurement geometry (horizontal, vertical, tilted). Mapping measurements have been performed on the gable "God the Father with Angels" by Giotto, employing a tube voltage of 50 kV, a tube anode current of 80 μA, and an acquisition d-well time of 1 s for each point. The total XRF map, 60 × 30 mm with a lateral step of 1 × 1 mm, has been recorded in about 50 min. The isolated XRF single-point spectra have been collected with slightly different acquisition parameters: acquisition time of 120 s, tube voltage between 20 and 50 kV, tube anode current between 40 and 200 μA depending on the point.

2.1.2. ARTAX 200 Spectrometer by Bruker

Bruker portable EDXRF spectrometer is equipped with an X-ray tube (Mo anode) with a beam collimated at 0.65 mm in diameter corresponding to an excited sample area of 0.33 mm^2. The characteristic X-ray radiation emitted by the sample is passed to a semi-conductor detector, which works according to the drift chamber principle (SDD). The idea of optical triangulation is applied for the adjustment of the working distance between the sample and the measuring head. The excitation beam and the optical axis of the detector meet at the measuring point when the laser spot matches cross hairs reported on the service CCD camera of the system; this procedure guarantees the reproducibility of the experimental conditions. The system presents an exchangeable filter slide with three positions and its sensitivity ranges from 2 to 40 keV. In this study, the Mo filter (12.5 μm thick) has been used, the tube voltage and the anode current of the tube have been fixed, respectively, to 40 kV and 800 μA, the acquisition time was 60 s while the acquisition window was 1–25 keV.

2.2. Data Handling and Samples

The idea behind the design of STEAM is to extend the information contained in a well-characterized area(s) of the sample outside its edge. In this perspective, it requires the region's data to be referred to as a starting point (i.e., a database) for investigating new spectra and the portions of the surface they belong to. Therefore, before the application of the protocol to uncharacterized works of art, the capabilities of STEAM were checked on an ad-hoc prepared sample.

2.2.1. The MA-XRF Database

The database is constituted by 1800 spectra mapped over a 60 × 30 mm area (yellow rectangle and zooms in Figure 1) of the surface of the gable of San Diego by the spectrometer ELIO.

This choice finds its rationale in the fact that the features of the database must constitute the starting point for STEAM, and therefore they must be well characterized. The area in Figure 1 has been exhaustively described by Galli and co-workers [46]; therefore, its main features can be considered as known. For what concerns the application of our protocol, the main results of the quoted paper can be summarized as follow: the results referring to the left half of the scanned area mainly compliant with the technique used by Giotto in the Croce of Santa Maria Novella, a double layer providing a Cinnabar-White Lead mixture (elementally dominated by the couple Pb-Hg) superimposed to a gypsum-Verdaccio lower layer (Ca-Sr and Fe-K couples) [47]. On the contrary, the right half reveals heavy restorations characterized by the presence of Ti and Cr; nothing can be said about the stratigraphy within the restoration areas other than the original layers have been lost or, at best, altered. The greenish tone of the infra-red false-color image (right zoom in Figure 1) highlights the original work by Giotto from the principal retouches which, for clarity, have been segmented and surrounded by red profiles.

2.2.2. The Reference Panel

Table 1 summarizes the ad-hoc layer sequences used to test the capabilities of STEAM. All of them were made with pure red pigments spread in egg tempera or oil on a panel board prepared with White Lead over gypsum. The mockup paintings, drawn in separate areas, are made of a few layers combined in different ways to verify if the method was able to discern within similar, but different, cases. Pigment choice has been based on the palette used by Giotto for the flesh tones of the gable of San Diego [15,46]. In order to account for the irregular thickness of the painted panel, three XRF spectra have been acquired by Bruker Artax 200 spectrometer for each layers' sequence and the average spectra have been used to test the performance of the protocol.

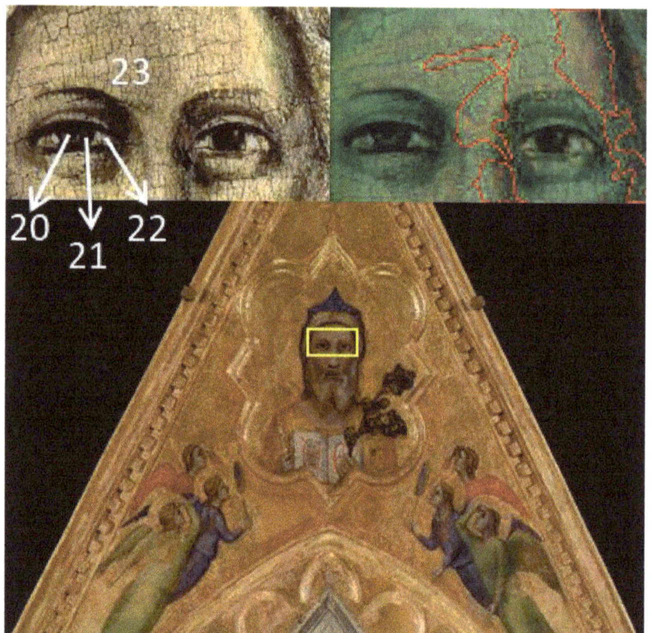

Figure 1. The gable of San Diego; the yellow rectangle surrounds the area of the X-ray fluorescence (XRF)-mapping used as a database. The Left Zoom of the scanned area highlights the locations of the isolated spectra collected within the mapped zone; the numbers refer to Table 2. The Right Zoom shows the infrared false-color image of the scan: the greenish tone reveals the original layers while the red profiles outline the principal zones affected by restorations.

2.2.3. Punctual XRF Test Set

Twenty-three isolated XRF spectra have been collected by ELIO on the gable of San Diego, inside and outside the area considered for MA-XRF scanning, as detailed later. The location of the acquisition points (Figure 2 and left zoom in Figure 1) has been chosen by art historians on the basis of the available documentation about the conservation history of the painting.

Table 1. Sequences of the pigments used in the reference panel. Every sequence is reported from the bottom to the upper layer. Reference numbers used in results are indicated in brackets.

■ Red Lake(oil) (1)	■ Cinnabar (oil) (2)	■ Hematite (oil) (3)	■ Red Ocrhe (oil) (4)
■ Carmine (oil) (5)	■ Carmine (tempera) (6)	■ Red Ocrhe (tempera) ■ Cinnabar (tempera) (7)	■ Cinnabar (tempera) ■ Red Ocrhe (tempera) (8)
■ Red Lake (tempera) (9)	■ Cinnabar (tempera) (10)	■ Red Ocrhe (tempera) (11)	■ Hematite (tempera) (12)
■ Cinnabar (tempera) ■ Red Lake (tempera) (13)	■ Red Ocrhe (tempera) ■ Red Lake (tempera) (14)	■ Cinnabar with White Lead (tempera) (15)	■ Cinnabar, Red Lake with White Lead (tempera) ■ Red Lake (tempera) (16)
■ Hematite with White Lead (tempera) ■ Red Lake (tempera) (17)	■ Cinnabar, Red Lake with White Lead (oil) ■ Red Lake (oil) (18)	■ Red Lake, Cinnabar with White Lead (tempera) (19)	■ Verderame, Yellow Lead, Red Lake with White Lead (tempera) ■ Red Lake (tempera) (20)

Figure 2. Location of the isolated XRF spectra collected outside the mapped area. The numbers refer to Table 2.

The choice of these points follows the twofold purpose to verify hypotheses from previous studies [15,46,48] and, if possible, to gather new insights into the gable's features. Since the right part was supposed to be more affected by restorations with respect to the left one, the same tones have been collected on both groups of angels to compare the used pigments. Moreover, the book held by God has been investigated because the authenticity of the letters typed on the pages is still debated. The details of the points where the spectra have been detected are listed in Table 2 and can be visualized in Figures 1 and 2. Points from 20 to 23 have been collected within the scanned region to guarantee a connection between the punctual and mapping XRF acquisition in this validation phase.

Table 2. Points of the gable of San Diego where the punctual XRF spectra have been collected. The numbers between the parentheses refer to Figure 1 (from 20 to 23) and Figure 2 (1 to 19).

Right Angels Green dress (1)	Right Angels Blue dress (2)	Right Angels Face of the green dressed (3)	Right Angels Hair of the green dressed (4)	Right Angels Halo of the green dressed (5)
Right Angels Sleeve of the green dressed (6)	Right Angels Red dress (7)	Right Angels Hair of the blue dressed (8)	Left Angels Green dress (9)	Left Angels Blue dress (10)
Left Angels Sleeve of the bleu dressed (11)	Left Angels Red dress (12)	Left Angels Face of the red dressed (13)	Book Black Ink (14)	Book White of the pages (15)
Book Red ink (16)	God the Father Yellow dress (17)	God the Father Blue hat I (18)	God the Father Blue hat II (19)	God the Father Pupil (20)
God the father Iris (21)	God the Father Sclera (22)	God the Father Face (23)		

2.3. The STEAM Protocol

An X-ray fluorescence spectrum reflects the elemental composition of the materials used in the painting in the volume around the acquisition point [49]. Organizing the acquired spectra into maps (XRF mapping) [15,46], thus, allows to retrieve the spatial distribution of the elements. Global Spectrum (GS) [15] can be a statistical tool for the analysis of the scanned area(s), as it highlights the statistical weight of each element in the dataset. It allows to evaluate how frequently a particular element is found in a group of spectra: the intensity of a channel of the GS represents how many times that channel brings information within the spectra, so how many times a particular fluorescence signal is detected in the dataset. All these inputs not only provide an exhaustive description of the original and restored regions (including the used pigments) of a work of art but, when supported by appropriate statistical tools, also permit inferences of the stratigraphy of the mapped area(s) [17,46,50,51]. Unluckily, the achieved knowledge remains confined within the edge of the scan(s) and no significant conclusions can be deduced regarding data from outside the mapped area(s). Applying our STEAM protocol, the evidence retrieved from the scanned area(s) create the database, which can be employed to characterize the spectra outside the mapped region(s), even belonging to a different sample. Thus, the XRF study of the objects of interest can be initially performed without the need for a complete scan of the sample(s)' surface; indeed, the MA-XRF can be planned in a second time and tailored on the base of the results of STEAM on the single points spectra.

STEAM protocol was developed exploiting Matlab programming (MATLAB and Statistic Toolbox Release 2019b, The MathWorks, Inc., Natick, MA, USA) with the aim of becoming an automatic process useful also to non-expert users.

2.3.1. STEAM vs. SAMs

As evident by the acronym, STEAM is inspired by SAM [43]. As mappers, the protocol processes X-ray spectra as N-dimensional vectors, where N is the number of acquisition channels of the spectrometer and assesses the similarity between a specific spectrum, ρ_k, and the $\rho'_{(i,j)}$ spectra belonging to the mapped region(s) by calculating the angle $\alpha_{(i,j)_k}$ between the spectra, that, accordingly to Equation (2), is:

$$\alpha_{(i,j)_k} = arccos\left(\frac{\sum_{l=1}^{N}\rho_{k_l}\rho'_{(i,j)_l}}{L_{\rho_k}L_{\rho'_{(i,j)}}}\right) \tag{3}$$

where:

$$L_{\rho_k} = \sqrt{\sum_{l=1}^{N} \rho_{k_l}^2} \tag{4}$$

and

$$L_{\rho'_{(i,j)}} = \sqrt{\sum_{l=1}^{N} \rho'^2_{(i,j)_l}} \tag{5}$$

are the lengths of the vectors, and ρ_{k_l}, $\rho'_{(i,j)_l}$, are the counts at the l-th channel of the spectrometer (i.e., the l-th coordinate of the vectors). In SAMs, L_{ρ_k} refers to the length of the k-th endmember where k is an integer between one and the number of identified endmembers; i.e., the idealized members of the elements' classes recognized within the data [52]. The task of SAMs is to reformulate the $\rho'_{(i,j)}$ as linear combinations of ρ_k with coefficients dependent on the angles $\alpha_{(i,j)_k}$. When mapping refers to paintings, the set of ρ_k is a pool of independent spectra identified by the user as the best descriptors of the investigated palette [53], while the $\rho'_{(i,j)}$ are MA-XRF acquisition organized into a 3D spatial-energy matrix. Therefore, every endmember corresponds to a so-called similarity map $[A]_k$, an image that describes the spatial affinity of the k-th endmember with the spectra within the scanned region(s) [54]. The lower is the angle $\alpha_{(i,j)_k}$ the higher is the brightness of the image (i.e., of the matrix element $\alpha_{(i,j)_k}$ in $[A]_k$) and the higher is the similarity of the k-th endmember with that point of the scan. Since the endmembers are independent of each other, the $[A]_k$ identifies the map distribution of the pigments chosen as k-th endmembers for the dataset.

STEAM still evaluates angles between vectors in N dimensions, but it does not refer to endmembers; ρ_k are the spectra of k uncharacterized isolated points that can be equally acquired within the mapped zone, into another area of the same sample or, even, derived from other object(s) of interest. In this case, every $[A]_k$ still shows which portion(s) of the scan best matches the features of the associated ρ_k, but since the isolated spectra are not necessarily independent of each other, an exhaustive description of the $\rho'_{(i,j)}$ in terms of ρ_k or vice versa is not feasible. Indeed, the most intense channel shared with the k-th spectrum dominates every element of the matrices $[A]_k$: elements that are abundant and present in the whole map quite never correspond to the most enlightening ones; in fact, the peculiar chemical element for a pigment (hereafter referred as the "signature") of an artwork is often a minor component in the mixture that constitutes the applied layer(s). The small, but significant coordinates, ρ_{k_l}, bound to the signature, risks to be underrated or darkened by the higher ones. As an example, in the case of various colored pigments in a mixture with a high quantity of the same white pigment, the small quantity of chromophore pigment would be hidden from the high signal identifying the white used. The absence of endmembers seems to be a pitfall of STEAM; however, the developed protocol compensates for the lack of independent endmembers employing an innovative use of the statistical features describing the XRF spectra and summarized by the GS. Here, the role of ρ_k and $\rho'_{(i,j)}$ is switched as the map itself describes the isolated points, thus no effort for the search of descriptors should be done and the database potentially works for an unlimited number of datasets.

2.3.2. STEAM and GS

In the GS, the intensity of each fluorescence peak represents the frequency of observation of that peak within the dataset [15]. However, the global spectrum does not merely highlight the q main fluorescence lines within the dataset (where q ranges between 0 and the number of series gathered by the GS): it is easy to build a hierarchy within the GS, based on the rational hypothesis that the channels characterized by the highest counts will be barely meaningful for classification purposes since a high intensity within the GS means that the corresponding fluorescence line has been detected in quite all the spectra. Therefore, the related element will result useless for inferring the significant features of the regions surrounding the isolated points (i.e., these elements do not belong to the signature).

The Statistically Tailored Elemental Angle Mapper translates the statistical abundance into a criterion for manipulating the spectra before the evaluation of $\alpha_{(i,j)_k}$: the components ρ_{k_l} and $\rho'_{(i,j)_l}$ (corresponding to the ranked channels) are sequentially deleted, starting from the most recurrent, in this way the role of the low rated components is emphasized. Thus every ρ_k is associated with a set of q similarity maps, $[A]_{k_q}$. The value of q represents both the place hold by the fluorescence line within the ranking and the order of deletion ($q = 0$ means that no channel has been removed). The abundance has been assumed as the first ranking mode, indeed it is the more direct statistical parameter a ranking criterion can be based on. In fact, the deleting order depends on the counts in the GS and the rationale is straightforward to understand: a fluorescence peak detected in almost all the spectra cannot be effective for discerning between similar spectra and should be removed as first. Besides the abundance, STEAM can be tuned on a second-ranking mode defined as the incidence criterion: the fluorescence peaks have been ordered on the basis of how much their removal would affect the similarity distributions. The method compares the differences between $[A]_{k_0}$, the set of similarity maps obtained by the original spectra, and the possible q $[A]_{k_1}$, the sets of similarity maps obtained after that one of the q detected fluorescence peaks have been removed from the original spectra. Then, the brightness change (corresponding to a similarity change in STEAM) of the maps is evaluated as the number of pixels undergoing a brightness change bigger than the 10%. The fluorescence peak that corresponds to the maximum brightness change is ranked as the first of the removal list, the second place is assigned to the series that records the second biggest numbers of matrix elements that present a brightness change of 10% of their original value and so on. The rationale of the incidence criterion relies on the fact that the bigger is the number of scanned area's points affected, the lesser is the distinguishing power of a fluorescence peak.

Note that no new concepts need to be invoked for understanding STEAM. Every $[A]_{k_q}$ remains a set of similarity maps obtained evaluating the angle between couples of vectors, the only difference within the $[A]_{k_q}$ relies on the fact that the length of the vectors used for the evaluation of the $\alpha_{(i,j)_{k_q}}$ decreases at the increasing of q, due to the removal of a fluorescence peak at each step. A critical observation of the resulting sets of similarity maps potentially enables the users to exploit the database for inferring a description of the new data points and their local surrounding regions; the only sure exception is constituted by those isolated acquisitions situated at the dividing lines between two distinct areas of the sample (e.g., the border between two shades of paint); in this case, the description of the nearby region has to be considered as partial. Figure 3 shows a diagram that resumes as STEAM works.

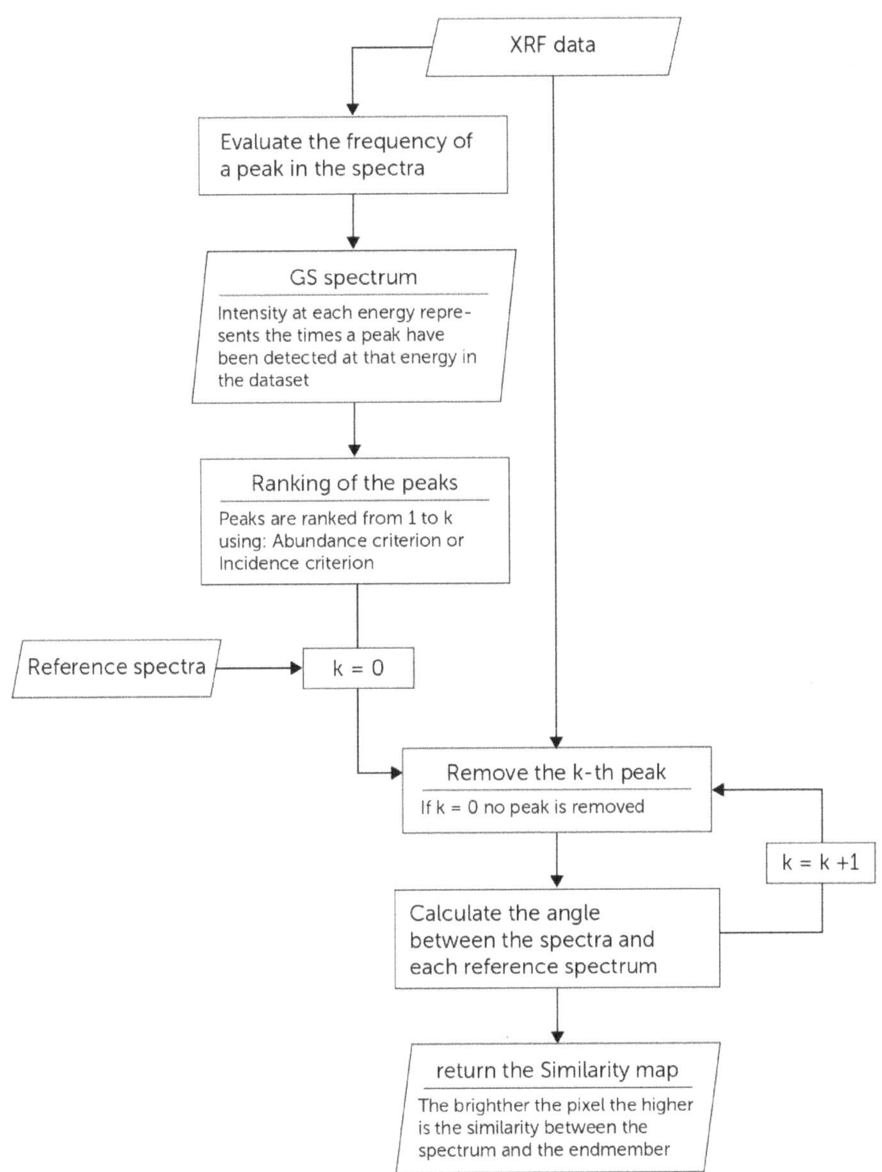

Figure 3. Statistically Tailored Elemental Angle Mapper (STEAM) flow chart.

3. Results

3.1. Characterization of the Reference Panel

The capabilities of STEAM have been tested by applying the protocol to a panel containing a set of samples whose characteristics were known (see Table 1). Table 3 reports the series emerging from the GS obtained from the spectra of the reference panel and ranked following the abundance criterion. The list establishes the order of removal for the series (ranking equal to zero, $q = 0$, means that no channels have been deleted from the spectra before the evaluation of the corresponding set of similarity maps, $[A]_{k_0}$).

Table 3. The principal energies (i.e., elements' emission series) emerging from the GS relating to the reference panel data. The series has been ordered following their abundance from the most to the less detected. The energy of the emission has been reported in keV and classified according to the X-ray nomenclature. The letters refer to Figure 4 and specify the series that have been removed before the calculation of the corresponding set of similarity maps.

Panel	A		B	C			D			
Ranking (q)	0	1	2	3	4	5	6	7	8	9
Energy (keV)		10.5	12.6	9.2	13.8	8.8	6.4	11.8	10.0	11.3
X-ray transition		L_α (Pb)	L_β (Pb)	L_I (Pb)	L_γ (Hg)	L_I (Hg)	K_α (Fe)	L_β (Hg)	L_α (Hg)	L_η (Pb)

The first places of the ranking are a prerogative of the series of the Pb and this is straightforward: Lead should appear in nearly all the analyzed spectra since the pigments' sequences have been spread on a panel prepared with a layer of White Lead over gypsum. On the other hand, the peaks related to the X-ray lines of elements such as Hg (L_α at 10.0 and L_β at 11.8 keV) or Fe (K_α at 6.4 keV), which characterize, respectively, cinnabar and the couple hematite/red ochre (red ochre is a clay colored by a variable amount of hematite), come toward the end of the list because their contributions have been detected in a limited number of sequences.

Figure 4 shows the sets of similarity maps, $[A]_{k_q}$, resulting from the comparison between the spectra of the database (the Giotto's gable) and those collected on the reference panel. Figure 4A shows the results obtained without manipulating the spectra. It is possible to observe the presence of different groups of similarity distributions within $[A]_{k_0}$, but any partition is easily ascribable to the features reported in Table 1. Being hematite/red ochre and red lake pigments with very different composition, the sequences they belong should be distinguishable, but this does not arise from $[A]_{k_0}$ even if hematite/red ochre is characterized by Fe while red lake does not contain elements detectable by either the spectrometers. Cinnabar-based sequences cannot be clearly recognized too; even if Hg should be a distinctive element, from Figure 4 it is quite difficult to suppose a significant difference between the sequences 15 (Cinnabar) and 5 (Carmine/Red Lake) or argue that 15 and 10 both contain Cinnabar (Hg). The fact that Hg and Fe do not constitute a peculiar feature is not due to the low sensibility of the instruments, otherwise, these elements would not be noticed within the global spectrum; it could be that the counts for the peaks at 6.4 (K_α emission of Fe), 10, and 11.8 keV (respectively, L_α and L_β emissions of Hg) are too low for making iron or mercury a distinctive characteristic of the related spectra. Essentially, Hg and Fe cannot be a real signature because mercury and iron's X-ray main series have low counts in comparison with Lead which, in turn, dominates and flattens the $[A]_{k_0}$ similarity maps as foregone since the panel has been prepared with White Lead. Figure 4B shows what happens not considering the 10.5 and 12.6 keV peaks (respectively, L_α and L_β emissions of Pb) in the data. Three principal groups appear and strikingly they fit the description of Table 1 quite at all: 2, 10, 13, 15, 16, 18, and 19 reflect the presence of cinnabar; 3, 11, 12, 14, and 17 individuate the hematite/red ochre group, while 1, 5, 6, 9, and 20 correspond to Red Lake. The remaining sequences, 4, 7, and 8, deserve further discussion. Sequence 4 is Red Ochre over White Lead, therefore, it should be characterized by Fe and resemble the similarity distribution of the Hematite/Red Ochre group. Probably, the intensity of K_α of Fe for sample 4 is particularly weak in comparison with the intensity of Pb's lines even if the principal lines in Pb L series have been already removed. This hypothesis finds a double confirmation in Figure 4C, obtained by deleting the 9.2 keV peak, Pb L_l line: firstly, the similarity distribution of sequence 4 approaches that of 3 and, in addition, the removal of 9.2 keV line also affects 11, 14, and 17, which contain red ochre too. It must be noticed that, after deleting the 3rd series of the ranking, a slight change can be observed for all the multi-layers as could be expected for a peak belonging to Pb. Even if $[A]_{k_3}$ do not suggest relations between sequences 7 and 8 (both containing Red Ochre; 7 below, 8 above Cinnabar) and the Hematite/Red Ochre group, the similarity maps of Figure 4C can be definitively grouped into 3 sets related, respectively, to Red Lake, Cinnabar and to the couple Hematite/Red Ochre. Once Iron has been eliminated ($[A]_{k_6}$, Figure 4D) Hematite/Red Ochre's group vanishes since it loses its peculiarity. The Red Lake set behaves in the same way probably because of the absence of Fe, a condition which marks this group and that, without the channels referred to Iron, stops to be significant. Summarizing, STEAM discerns three principal sets within the sequences of the reference panel dominated, respectively, by Hg, Fe, and the absence of Fe. Moreover, kept out the comments about 7 and 8, every consideration about the sequences of these samples' results is reliable also with any a priori knowledge of the reference panel' layers. The behavior of 7 and 8 is probably due to the shading action of Hg over Fe; thesis which finds prompt confirmation by manually switching the removal order for Iron and Mercury. If the 10 keV L_α and the 11.8 keV L_β of Hg are removed before the 6.4 keV K_α of Fe ($[A]_{k_{6*}}$, Figure 4E) the Cinnabar group vanishes, 7 and 8 reveal the presence of Fe, and their similarity maps can be associated with the Hematite/Red Ochre's group. Simultaneously, 2, 10, 13, 15, 16, 18, and 19 lose their signature and assimilate to the Red Lake's group consistently with the fact that Cinnabar itself does not contain Iron.

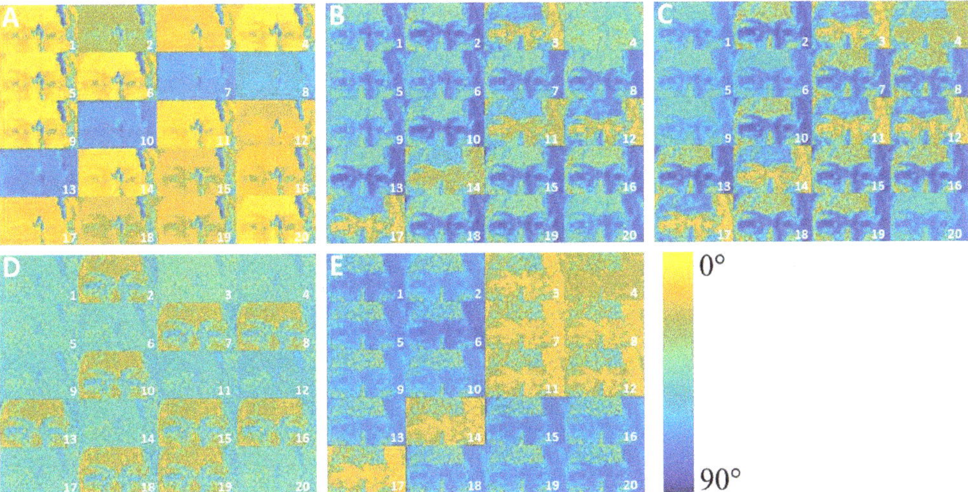

Figure 4. The sets of similarity maps, $[A]_{k_q}$, for the reference panel obtained following the abundance criterion. The numbers within the panels refer to Table 1, the letters to Table 3. The color bar illustrates the level of similarity (expressed in terms of the angle between vectors). (**A**) shows $[A]_{k_0}$, the set of similarity maps obtained without manipulating the spectra. (**B**) shows $[A]_{k_2}$, obtained after that the channels related to 10.5 and 12.6 keV have been suppressed. (**C**) shows $[A]_{k_3}$, where the 9.2 keV peak has been removed too. (**D**) shows $[A]_{k_6}$, the similarity maps after the removal of Fe related energies. (**E**), $[A]_{k_{6*}}$, shows what happens manually switching the order of removal for Hg and Fe.

The same conclusions can be inferred by changing the STEAM's ranking criterion from abundance to incidence.

Figure 5 shows what happens using as ranking's rationale the number of pixels involved by a significant similarity change (greater than the 10% in terms of image brightness) after the removal of one of the detected X-ray series. Again, the first positions of the ranking reflect the omnipresence of Lead while the signature elements, Hg and Fe, come later in the list (Table 4). Following the removal order reported in Table 4, it is straightforward that

Figure 5B,C, that refers to the similarity maps' sets $[A]_{k_2}$ and $[A]_{k_4}$, uncover the shading effect of Mercury over Iron with any ad-hoc manipulation of the ranking. Figure 5B confirms the splitting of the sequences into the Red Lake, Cinnabar, and Hematite/Red Ochre groups. Figure 5C unveils the presence of Iron simultaneously with Mercury for sequences 7 and 8. Analogously to the previous discussion, the Cinnabar's sequences do not contain Fe and, therefore, assimilate to Red Lake because of the lack of Fe. Since it is the most intuitive and the easiest to be implemented, the abundance criterion remains the first option for listing the removal order, but the discussion about sequences 7 and 8 proves the importance of exploiting both the available ranking criteria.

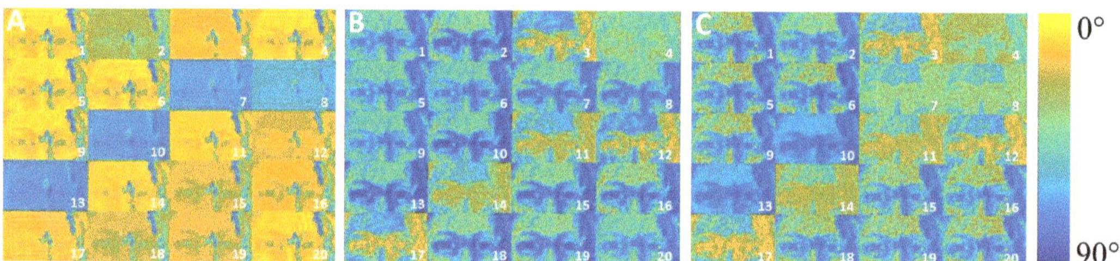

Figure 5. The sets of similarity maps, $[A]_{k_q}$, for the reference panel obtained following the incidence criterion. The numbers within the panels refer to Table 1, the letters to Table 4. The color bar illustrates the level of similarity (in terms of the angle between vectors). (**A**) shows $[A]_{k_0}$, the set of similarity maps obtained without manipulating the spectra. (**B**) shows $[A]_{k_2}$, obtained after that the channels around 12.6 and 10.5 keV have been suppressed. (**C**) shows $[A]_{k_4}$, the 11.8 and 10 keV series have been removed too.

Table 4. The principal energies (i.e., elements' emission series) emerging from the GS evaluated from the reference panel data. The series has been ordered following the incidence criterion: the first position is occupied by the series, which, if removed, generates a similarity change of more than 10% with respect to the value of $[A]_{k_0}$ involving the largest number of pixels; the second place is held by the series which involves the second larger number of pixels and so on. The energy of the emission has been reported in keV and classified according to the X-ray nomenclature. The letters refer to Figure 5 and specify the series that have been removed before the calculation of the corresponding set of similarity maps.

Panel	A		B		C				
Ranking (q)	0	1	2	3	4	5	6	7	8	9
Energy (keV)		10.5	12.6	11.8	10	6.4	9.2	11.3	13.8	8.8
X-ray transition		L_α (Pb)	L_β (Pb)	L_β (Hg)	L_α (Hg)	K_α (Fe)	L_I (Pb)	L_η (Pb)	L_γ (Hg)	L_I (Hg)

3.2. Characterization of the Punctual XRF Test Set from the Gable "God the Father with Angels"

Once the effectiveness of both the STEAM's ranking criteria has been demonstrated, the protocol has been applied to the case study of the isolated spectra of Giotto's gable "God the Father with Angels". In Table 5, the series has been listed following the abundance criterion and, therefore, they are ordered from the most to the less detected. Spectra from 20 to 23 have been chosen within the scanned area (see Figure 1) following a twofold rationale: on one hand, they should assure a link between the database and the isolated points; on the other, they should work as internal references for discussing the application of the statistically tailored elemental angle mapper.

Table 5. The principal energies (elements' emission series) emerging from the GS are evaluated from the isolated spectra. The series has been ordered following their abundance from the most to the less detected. The energy of the emission has been reported in keV and classified according to the X-ray nomenclature. The letters refer to Figure 6 and specify the series that has been removed before the calculation of the corresponding set of similarity maps.

Panel	A	B			C	D	E	F	G		H		I	
Ranking (q)	0	1	2	3	4	5	6	7	8	9	10	11	12	13
Energy (keV)		10.5	4	14.1	3.6	6.4	12.6	4.5	10	3.3	7	4.9	9.1	6.9
X-ray transition		L_α (Pb)	K_β (Ca)	K_α (Sr)	K_α (Ca)	K_α (Fe)	L_β (Pb)	K_α (Ti)	L_α (Hg)	K_α (K)	K_β (Fe)	K_β (Ti)	L_I (Pb)	K_α (Co)

The first positions of the ranking reflect the elements indicated by Galli and co-workers as the main components of the two layers used by Giotto: Pb (L_α at 10.5 keV), Ca (3.7 and 4 keV refer, respectively, to the k_α and the K_β of Ca) and Sr (K_α at 14.1 keV) [15,46]. The X-ray lines of Fe, Hg, and Ti do not compare in all the spectra and, analogously to the case of the reference panel (Table 3), they occupy lower places in the list of Table 5. Despite the stratigraphy below the isolated spectra is not available, STEAM is still able to exploit the database for unveiling significant features of the local areas the points belong.

Figure 6A depicts the similarity maps calculated without manipulating the spectra. Some differences can be easily observed, but it is quite difficult to classify the spectra referring to the appearance of the painting (see Table 2): neither the tones nor the locations or whatever self-evidence could drive the user toward a reliable data assignment. Once the peak at 10.5 keV (L_α of Pb) has been deleted (Figure 6B), no sensational improvements can be observed, but some maps show little uneven regions (purple arrows in Figure 6), not noticeable in the $[A]_{k_0}$ set, which will become important hereafter. $[A]_{k_4}$ (Figure 6C) are the key turn for the analysis of this group of isolated points; the 4th set of similarity maps shows how the scenario promptly clears up after the removal of Ca peaks at 3.7 (K_α) and 4 keV (K_β) and Sr peak at 14.1 keV (K_α). The couple Ca-Sr and Pb are probably the main components of the preparation layers and, once they have been deleted, the elements that define the tones of the pigments remain. Figure 6C suggests the first classification by means of the database's features graphically resumed in Figure 1. For clarity, it should be reminded that the red edged areas in Figure 1 have been obtained segmenting the average projection of the normalized elemental maps of Ti and Cr and, therefore, these regions refer to recent intervention(s) [15]. Since they show very low or even no similarity with the areas ascribed to the original palette, some spectra within the isolated points can be assigned to restoration event(s), such as numbers 2, corresponding to the blue dress of the angel at the right side, and 19, from the blue hat. Other points are consistent with the regions assigned to Giotto's work and can be considered as part of the original project of the master (points 10, 11, or 20, for instance). At this step of data elaboration, nothing can still be said about a few points: 5 and 7 seem to depict an impossible mix of original and modern pigments while, if on one hand 12 and 16 can be excluded from restorations, on the other hand, their similarity with the original areas is so low that any kind of assignment would be a hazard. Following the removal of the K_α line of Fe (6.4 keV, Figure 6D), the distinction between original and restoration areas results improved: 1, 2, 3, 4, 6, 8, 9, and 19 can be assigned to recent intervention(s) while 10, 11, 13, 14, 15, 17, 18, 20, 21, 22, and 23 to the Giotto's palette. 12 and 16 are probably two different kinds of reds ascribable to the Florentine master, but their characterization cannot be fully accomplished using the available database. The remaining red tone, 7, gives up some doubts about its collocation within the original pigments and, finally, nothing can still be inferred about point 5. After the removal of the 12.6 keV peak (L_β of Pb, Figure 6E), deductions from Figure 6D are confirmed and a further classification within the supposed original spectra is suggested. Areas around points 20 and 21 are characterized by Fe. Spectra 13, 14, 15, 22, 23 and their surrounding area can be exhaustively described in terms of the pigments identified by Galli and co-workers [15,46]: 13 and 23 as flash tones; 15, 22, and 14 in terms of the preparation layers because whites have no hues and the lines related to the black ink emissions are not detectable in the used energy range. The classification of 10, 11, 17, and 18 cannot go beyond their assignment to the original palette because blue or yellow hues do not appear in the database used. The low similarity for the maps of 12 and 16 suggests that both the mixtures used for these reds would employ the same materials used within the mapped region but mixed with different proportions. The removal of K_α line of Ti (4.5 keV) finally defines the classification of the isolated spectra (Figure 6F): the similarity within the maps of the points ascribed to the original palette increases in the same areas where it decreases for points assigned to restoration (red arrows in Figure 6) making this behavior a discriminant factor between Giotto and modern pigments. Even if the removal of the Ti series at 4.5 keV affects all the spectra, particularly eye catching is the behavior of some points, namely 2, 19, and 8. The first two completely lose the connection with the database confirming that neither restoration pigments can be fully characterized by this database alone. The similarity distribution of the latter is of great significance for spectra classification: it not only supports the need for expanding the database to get a full description of the pigments, but it also suggests the assignation of 7 to restoration because $[A]_{7_7}$ similarity map for point 7 resembles quite faithfully the features of the map related to 8, $[A]_{7_8}$. The deletion of L_α of Hg (Figure 6G) makes the original reds 12 and 16 lose their signature and become very similar to flash tones, vanishing one of the subsets identified by Figure 6D. Figure 6H follows the removal of the K_β line of Fe and, as by now expectable, the original spectra reduce into a single set while 7 can be finally associated with restoration(s). The definitive removal of Ti (Figure 6I) reveals another interesting aspect regarding the restoration areas. Modern pigments could be different even if they have been used to correct the same tone: 9 is not the same as 1 and 6 even if they have been employed to retouch the green dresses of the Angels. This suggests the presence of restorations made in different periods or performed by more than one restorer. Besides the classification of the data, STEAM promotes the discussion about the stratigraphy of the gable; in fact, the comparison between Figure 6A–C,E explains the small areas spotted by the purple arrows in Figure 6. The removal of the Pb L_α at 10.5 keV unveils the presence of uneven little areas in Figure 6B where the region was homogeneous in Figure 6A. Once removed, the K_α and the K_β of Calcium and the K_α of the Sr, the same regions disappear (Figure 6C) for being back after the removal of the L_β of Pb (Figure 6E). It seems quite evident that the contributions of Pb hide the couple Ca-Sr supporting the idea of a two-layers stratigraphy for the scanned zone, where the first containing Pb lies over the Ca-Sr-based one as argued by Galli and co-workers [46]. Switching from the abundance to the incidence criterion (see Table 6 for the list of the series), it is possible to support and extend the observation about the isolated point's dataset.

Table 6. The principal energies (elements' emission series) emerging from the GS evaluated from the isolated spectra. The series has been ordered following the incidence criterion: the first position is occupied by the series, which, if removed, generates a similarity change of more than 10% with respect to the value of $[A]_{k_0}$ involving the largest number of pixels; the second place is held by the series, which involves the second largest number of pixels and so on. The energy of the emission has been reported in keV and classified according to the X-ray nomenclature. The letters refer to Figure 7 and specify the series that have been removed before the calculation of the corresponding set of similarity maps.

Panel	A	B		C				D	E					
Ranking (q)	0	1	2	3	4	5	6	7	8	9	10	11	12	13
Energy (keV)		10.5	12.6	4	3.6	9.1	14.1	3.3	6.4	7	4.8	4.4	10	6.9
X-ray transition		L_α (Pb)	L_β (Pb)	K_β (Ca)	K_α (Ca)	L_I (Pb)	K_α (Sr)	K_α (K)	K_α (Fe)	K_β (Fe)	K_β (Ti)	K_α (Ti)	L_α (Hg)	K_α (Co)

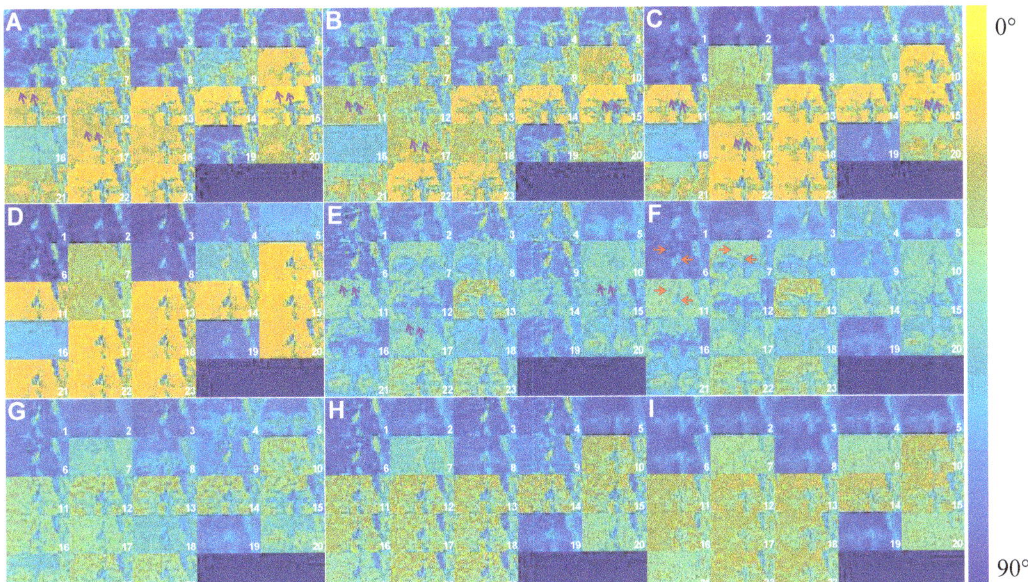

Figure 6. The sets of similarity maps, $[A]_{k_q}$, for the isolated test spectra of the Gable of San Diego obtained following the abundance criterion. The numbers within the panels refer to Table 2, the letters to Table 5. The color bar illustrates the level of similarity (in terms of the angle between vectors). (**A**) shows $[A]_{k_0}$, the set of similarity maps obtained without manipulating the spectra. (**B**) shows $[A]_{k_1}$, obtained after the channels related to 10.5 keV series of Pb has been suppressed. (**C**) shows $[A]_{k_4}$ which depicts what happens on removing the Ca series at 3.7 and 4 keV and the Sr series at 14.1. (**D**) shows $[A]_{k_5}$ that follows the removal of Fe. (**E**) shows $[A]_{k_6}$, the similarity maps after the removal of the 12.6 keV series of Pb. (**F**), $[A]_{k_7}$, follows the removal of the 4.4 series of Ti. (**G**) reports $[A]_{k_8}$, the similarity maps after the removal of the Hg series at 10 keV. (**H**), $[A]_{k_{10}}$, shows what happens if Fe is completely deleted. (**I**), $[A]_{k_{11}}$, corresponds to the removal of the residual Titanium series. Purple arrows in **A**–**C** and **E** and red arrows in (**F**) highlight the areas of interest discussed in the text.

Figure 7B promptly confirms the assignment of sequence 12 and 16 to the original palette of Giotto even if the mixtures used for these reds instantly appear different from the one in the scanned area. The sequential removal of Pb, Ca, and Sr recognizes red 7 as part of the restoration(s) (Figure 7C) basing on the observation of the areas identified by the red arrows (confirming what is inferred discussing Figure 6). $[A]_{k_7}$ (Figure 7D) marks the suppression of all the low energy contributions (minor or equal to 3.3 keV corresponding to the K_α emission of K) and allows to classify the couples formed by 11 and 18 and by 20 and 21, respectively, as original blues and blacks. The scenario for points 13, 14, 15, 17, 22, and 23 is only apparently more complex with respect to the situation depicted by the abundance criterion; once removed the Fe K_α series at 6.4 keV (Figure 7E), what we inferred observing the panels of Figure 6 can be proposed again. The agreement between abundance and incidence criterion is broken only by sequences 5 and 10 which assignment as unclassifiable and original switches, respectively, to original and restorations. The uncertainty about the nature of these two isolated points stresses once again the need for a larger database for achieving a full description for all the isolated spectra. Besides the data's assignment, the incidence criterion also confirms the mutual relationship between Pb and the couple Ca-Sr: the purple arrows in Figure 7 again highlights how the removal of Pb (Figure 7B) unveils the presence of an underlying layer basically constituted by Ca-Sr and linked to Fe detection (Figure 7C–E).

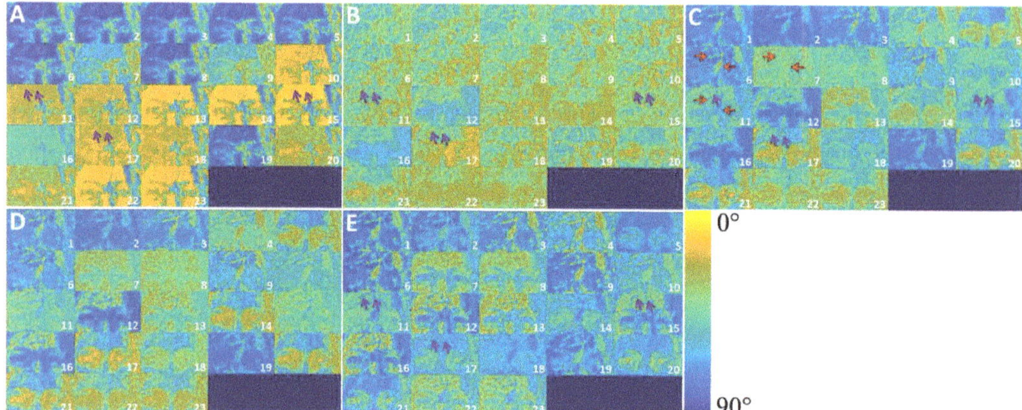

Figure 7. The sets of similarity maps, $[A]_{k_q}$, for the isolated test spectra of the Gable of San Diego obtained following the incidence criterion. The numbers within the panels refer to Table 2, the letters to Table 6. The color bar illustrates the level of similarity (in terms of the angle between vectors). (**A**) shows $[A]_{k_0}$, the set of similarity maps obtained without manipulating the spectra. (**B**) shows $[A]_{k_2}$, obtained after that the channels around 12.6 and 10.5 keV have been suppressed. (**C**) shows $[A]_{k_4}$, the kα and the kβ series of Ca at 3.7 and 4 keV have been removed. (**D**) shows $[A]_{k_7}$, the contributions minor or equal to 3.3 keV have been all removed. (**E**) shows $[A]_{k_8}$, the similarity maps obtained after the removal of the 6.4 keV kα series of Iron. Purple arrows in (**A**–**C** and **E**) and red arrows in (**C**) highlight the areas of interest discussed in the text (in (**C**) one of the purple arrows on point 11 has not been reported in order to not interfere with the details highlighted by the red arrows).

Excluding points number 5 and 10, the STEAM splits the isolated spectra dataset into two main groups: restoration(s) and original pigments which can be further divided into subsets depending on how their features match those of the database. The method solves, at least partially, ambiguous assignments, such as those of 7, 12, or 16 and, finally, it is a useful support in the determination of the stratigraphy. The removal rankings must be intended as deduced starting from the GSs of the datasets to be studied but nothing forbids to reformulate the removal lists employing the database's GS or, even the global spectrum obtained by the combination of both the GSs; obviously, if combined the two global spectra have to be normalized to the number of spectra they have been calculated from, otherwise abundance criterion cannot be considered significant.

4. Discussion

The application of STEAM to the reference panel widely tested the capabilities of the protocol for recovering the main features of the sequences. Starting from an unclear situation (Figures 4A and 5A) the protocol gradually reaches a description of the reference panel coherent with Table 1 (Figure 4C or Figure 5B). The partial lack of information regarding the role of Fe in sequences 7 and 8 can be overcome by exploiting the intriguing observation that the K_α of Fe, the L_β of Hg, and the L_α of Hg are ranked one after the other. This evidence suggests changing the removal order of Table 3 and, indeed, the switching of Fe with Hg promptly unveils the presence of Red Ochre in sequences 7 and 8 (Figure 4D,E) allowing the user to associate them to the Hematite/Red Ochre group too. If a manual change driven by a practical observation can be charged to be a discretional choice to the operator, changing the STEAM ranking mode to the incidence criterion returns a removal order (Table 4), which not only confirms the conclusions provided by the abundance criterion (Figure 5B) but also automatically elicits the similarity distribution of Fe within 7 and 8 (Figure 5C). Since there is nothing to prevent the sequential or simultaneous use of the two criteria, the capability of changing the ranking mode enhances the effectiveness of STEAM suggesting the user always reviews both the resulting sets of similarity maps. The application of the protocol to the case of 23 punctual spectra collected from different zones of the gable of San Diego (Figures 1 and 2) illustrates the STEAM's capabilities for a much more complex experimental situation with respect to the reference panel. The discussion of the panels of Figures 6 and 7 clarifies how the systematic use of both the ranking criteria of STEAM describes the features of the isolated points. The protocol reliably classifies 21 of 23 spectra into two main groups, original and restoration(s) pigments, which in turn can be divided into sub-groups on the base of their signature elements. It would be inappropriate to consider the points' description as complete because the database at our disposal was too small (see Figure 1 to qualitatively compare the dimension of the scanned area with those of the gable) to account for all the significant pigments of the sample, however, a classification of the spectra coherent with Table 2 is achieved (Panel F and E in Figures 6 and 7, respectively). The uncertainties about the assignment of points 5 and 10, or the lack of some features for other spectra (2, 7, 12, or 16 for example) cannot be down-listed as mere fails of the protocol but, rather, considered as areas of the sample worthy of further investigation and at least for new measurements' campaigns. Besides the description of the isolated points and their surrounding regions in terms of main elements' spatial distributions, the application of the protocol to the gable of San Diego proves how STEAM can support the debate about other characteristics of the investigated item: in this case, the protocol fosters the discussion on a crucial issue such as the stratigraphy of the scanned area. A careful evaluation of the sets of similarity maps sheds light on the mutual relationship between Pb and the couple Ca-Sr, the behavior spotted within the small areas pointed out by the purple arrows in Figures 6 and 7 provides confirmation for the hypothesis of stratigraphy given by Galli and co-workers [46]: two superimposed layers where the lower is characterized by Ca-Sr while the upper by Pb.

Finally, we notice that, in both application the panel and the gable cases, the abundance criterion shows an unexpected behavior concerning the ranking of the series. The abundances of Hg L_γ in Table 3 and of Ca K_β in Table 5 are greater than those of the principal lines of their series, namely Hg L_α and L_β, and Ca K_α, respectively. This is probably due to the low number (i.e., low statistic) of spectra (respectively, 20 for the panel and 23 for the gable) used for evaluating the GSs. The low statistic, together with the weakness of the counts within the energy ranges of the Hg L_γ and of the Ca K_β, can generate false positives (for spectra from painted regions without Hg or Ca) or missed peaks (for spectra from painted regions containing Hg or Ca) at the energies corresponding to these two series. Considering 20 spectra, a single false-positive or missed peak alters not only the GS but also the evaluation of the abundance for at least the 5% and, consequently, changes the positions of the relative series in the rankings. In general, this issue would deserve a deepen discussion; luckily, the un-expected ranking position for both Hg L_γ and Ca K_β affect none of the classification steps of the experimental data, as no panel in Figure 4 or Figure 6 deals with Hg L_γ or Ca K_β. We thus consider this issue of low relevance in the present case.

5. Conclusions

The information returned by macro-X-ray fluorescence mapping is identified with the spatial distributions of the main elements in the region of the scan and, if SAM's analysis is available, with the description of the spectra as linear combinations of the endmembers. In this general case, the obtained results help to extract the features of the mapped region and, with the support of other experimental and/or statistical methods, the palette of the master, the presence of restorations, or even the stratigraphy of the painting can be inferred with great benefit for both art historians and scientists. The main limitation is that what can be concluded remains constrained within the edges of the scanned area. The Statistically Tailored Elemental Angle Mapper is born in the attempt of extending the knowledge returned by p-MA-XRF imaging outside the acquisition's area(s). In the contest of STEAM, the mapped region becomes a database operating as a starting point for exploring the properties of new sets of data. Starting from the statistical picture given by the global spectrum,

STEAM distinguishes the significant series, orders the emission peaks through a suitable criterion, creates a hierarchy, and finally manipulates the spectra enabling the user to extract information evaluating sets of similarity maps. The test performed on the reference panel demonstrates the capabilities of STEAM. Starting from a scenario flattened by the emissions of Pb, the protocol gradually unveils the peculiarities of the sequences until the elemental description of the panel substantially reproduces the properties of the samples. The discussion on the multilayers containing both Red Ochre and Cinnabar shows how the availability of different ranking criteria is a key-turn rather than an optional for the effectiveness of the protocol. The subsequent application of STEAM to the sparse spectra from the gable of San Diego confirms that, even if the mapped region(s) cannot account for all the significant pigments, the protocol reliably classifies most of the isolated points providing a base for discussing and designing a new strategy for the characterization of the sample. In this case, the non-complete answers are due to a lack of a complete and specific spectra database. Besides the description of the XRF spectra, the application to the gable again shows how STEAM can be successfully involved in the determination of a complex issue such as the stratigraphy of the scanned area: the results seem to confirm the hypothesis given by Galli and co-workers [46]. In the shed of these considerations, STEAM can be adopted as an effective tool for extending the information from XRF map(s) outside the narrow edges of the scan(s); moreover, the statistically driven manipulation of the spectra not only points out peculiar details but also can take part in the discussion of several issues related to the experimental system. Surely, the protocol finds its main cue in the pioneering way employed for handling the spectra, but STEAM introduces also other significant novelties. The protocol is independent of the instrument used for acquiring the data; in fact, even if they have been simultaneously used, the spectra from the gable and from the reference panel have been obtained by two different spectrometers. Every database can be employed any time a new characterization is required; it is straight-forward that a database can be extensively useful only if the uncharacterized spectra and the database share their principal chemical elements but, also in the unlucky case in which this does not occur, STEAM can even enable the user to exclude some hypotheses in favor of others with an obvious advantage for the time and design of the experimental and analysis campaign. Finally, the possibility of applying multiple criteria gives an essential boost to the power of the protocol: even if the properties of a sample can be considered as largely known, the exclusion of one of the criteria could lower the effectiveness of the analysis as shown discussing the properties of the reference panel. In summary, the statistically tailored elemental angle mapper provides the first attempt to answer one of the new needs of applied X-ray fluorescence: a generalized way for exploiting the big amount of data made available by MA-XRF portable instrumentation. STEAM leaves aside the interest for the single case study or the taste for technological development for introducing an original statistical data handling method. STEAM has been proven to be effective but, at the same time, it has been designed to be always improved by increasing the database with the addictions of even more maps and updating the number or modality of the ranking criteria; therefore, the statistically tailored elemental angle mapper candidates for being the benchmark for the whole p-MA-XRF imaging community.

Author Contributions: Conceptualization, A.G., L.B., and M.C.; methodology, A.G. and M.C.; software, M.C.; validation, A.G. and M.C.; investigation, A.G. and L.B.; writing—original draft preparation, J.O., A.G., L.B., and M.C.; writing—review and editing, J.O., A.G., L.B., and M.C.; supervision, A.G. and L.B. All authors have read and agreed to the published version of the manuscript.

Funding: This research was funded by Fondazione Cariplo, grant no. 2015-2293.

Institutional Review Board Statement: Not applicable.

Informed Consent Statement: Not applicable.

Acknowledgments: The authors would like to especially thank Michael A. Brown, Associate Curator of European Art at the San Diego Museum of Art. The spectrometer ELIO has been kindly supplied by XGLab srl, Bruker Nano Analytics Division. For this, we are very grateful to Roberto Alberti and coworkers.

Conflicts of Interest: The authors declare no conflict of interest. The funders had no role in the design of the study; in the collection, analyses, or interpretation of data; in the writing of the manuscript, or in the decision to publish the results.

References

1. Tsuji, K.; Nakano, K.; Hayashi, H.; Hayashi, K.; Ro, C.-U. X-Ray Spectrometry. *Anal. Chem.* **2008**, *80*, 4421–4454. [CrossRef] [PubMed]
2. Romano, F.P.; Pappalardo, L.; Biondi, G.; Caliri, C.; Masini, N.; Rizzo, F.; Santos, H.C. FF-XRF, XRD, and PIXE for the Nondestructive Investigation of Archaeological Pigments. In *Sensing the Past: From Artifact to Historical Site*; Masini, N., Soldovieri, F., Eds.; Geotechnologies and the Environment; Springer International Publishing: Cham, Switzerland, 2017; pp. 325–336. ISBN 978-3-319-50518-3.
3. Romano, F.P.; Caliri, C.; Cosentino, L.; Gammino, S.; Giuntini, L.; Mascali, D.; Neri, L.; Pappalardo, L.; Rizzo, F.; Taccetti, F. Macro and Micro Full Field X-Ray Fluorescence with an X-Ray Pinhole Camera Presenting High Energy and High Spatial Resolution. *Anal. Chem.* **2014**, *86*, 10892–10899. [CrossRef] [PubMed]
4. Walter, P.; Sarrazin, P.; Gailhanou, M.; Hérouard, D.; Verney, A.; Blake, D. Full-field XRF Instrument for Cultural Heritage: Application to the Study of a Caillebotte Painting. *X Ray Spectrom.* **2019**, *48*, 274–281. [CrossRef]
5. Stromberg, J.M.; Van Loon, L.L.; Gordon, R.; Woll, A.; Feng, R.; Schumann, D.; Banerjee, N.R. Applications of Synchrotron X-Ray Techniques to Orogenic Gold Studies; Examples from the Timmins Gold Camp. *Ore Geol. Rev.* **2019**, *104*, 589–602. [CrossRef]
6. Pan, Y.; Hu, L.; Zhao, T. Applications of Chemical Imaging Techniques in Paleontology. *Natl. Sci. Rev.* **2019**, *6*, 1040–1053. [CrossRef]
7. Li, J.; Pei, R.; Teng, F.; Qiu, H.; Tagle, R.; Yan, Q.; Wang, Q.; Chu, X.; Xu, X. Micro-XRF Study of the Troodontid Dinosaur Jianianhualong Tengi Reveals New Biological and Taphonomical Signals. *bioRxiv* **2020**. [CrossRef]
8. Langstraat, K.; Knijnenberg, A.; Edelman, G.; van de Merwe, L.; van Loon, A.; Dik, J.; van Asten, A. Large Area Imaging of Forensic Evidence with MA-XRF. *Sci. Rep.* **2017**, *7*, 15056. [CrossRef]
9. Yan, J.; Chia, J.-C.; Sheng, H.; Jung, H.; Zavodna, T.-O.; Zhang, L.; Huang, R.; Jiao, C.; Craft, E.J.; Fei, Z.; et al. Arabidopsis Pollen Fertility Requires the Transcription Factors CITF1 and SPL7 That Regulate Copper Delivery to Anthers and Jasmonic Acid Synthesis. *Plant Cell* **2017**, *29*, 3012–3029. [CrossRef]
10. Lider, V.V. X-Ray Fluorescence Imaging. *Phys. Usp.* **2018**, *61*, 980. [CrossRef]
11. Romano, F.P.; Janssens, K. Preface to the Special Issue on: MA-XRF "Developments and Applications of Macro-XRF in Conservation, Art, and Archeology" (Trieste, Italy, 24 and 25 September 2017). *X Ray Spectrom.* **2019**, *48*, 249–250. [CrossRef]
12. Special Issue: First Workshop on Macro X-Ray Flourescence (MA-ARF) Scanning, 24 September 2017, Trieste, Italy. *X Ray Spectrom.* **2017**, *48*, 247–318. [CrossRef]
13. Alfeld, M.; Gonzalez, V.; van Loon, A. Data Intrinsic Correction for Working Distance Variations in MA-XRF of Historical Paintings Based on the Ar Signal. *X Ray Spectrom.* **2020**. [CrossRef]
14. Cavaleri, T.; Buscaglia, P.; Caliri, C.; Ferraris, E.; Nervo, M.; Romano, F.P. Below the Surface of the Coffin Lid of Neskhonsuennekhy in the Museo Egizio Collection. *X Ray Spectrom.* **2020**. [CrossRef]
15. Gargano, M.; Galli, A.; Bonizzoni, L.; Alberti, R.; Aresi, N.; Caccia, M.; Castiglioni, I.; Interlenghi, M.; Salvatore, C.; Ludwig, N.; et al. The Giotto's Workshop in the XXI Century: Looking inside the "God the Father with Angels" Gable. *J. Cult. Herit.* **2019**, *36*, 255–263. [CrossRef]
16. Alfeld, M.; Mulliez, M.; Devogelaere, J.; de Viguerie, L.; Jockey, P.; Walter, P. MA-XRF and Hyperspectral Reflectance Imaging for Visualizing Traces of Antique Polychromy on the Frieze of the Siphnian Treasury. *Microchem. J.* **2018**, *141*, 395–403. [CrossRef]
17. Uhlir, K.; Gironda, M.; Bombelli, L.; Eder, M.; Aresi, N.; Groschner, G.; Griesser, M. Rembrandt's *Old Woman Praying*, 1629/30: A Look below the Surface Using X-ray Fluorescence Mapping. *X Ray Spectrom.* **2019**, *48*, 293–302. [CrossRef]
18. D'Elia, E.; Buscaglia, P.; Piccirillo, A.; Picollo, M.; Casini, A.; Cucci, C.; Stefani, L.; Romano, F.P.; Caliri, C.; Gulmini, M. Macro X-Ray Fluorescence and VNIR Hyperspectral Imaging in the Investigation of Two Panels by Marco d'Oggiono. *Microchem. J.* **2020**, *154*, 104541. [CrossRef]
19. dos Santos, H.C.; Caliri, C.; Pappalardo, L.; Catalano, R.; Orlando, A.; Rizzo, F.; Romano, F.P. Real-Time MA-XRF Imaging Spectroscopy of the Virgin with the Child Painted by Antonello de Saliba in 1497. *Microchem. J.* **2018**, *140*, 96–104. [CrossRef]
20. Alfeld, M.; Pedetti, S.; Martinez, P.; Walter, P. Joint Data Treatment for Vis–NIR Reflectance Imaging Spectroscopy and XRF Imaging Acquired in the Theban Necropolis in Egypt by Data Fusion and t-SNE. *Comptes Rendus Phys.* **2018**, *19*, 625–635. [CrossRef]
21. Galli, A.; Gargano, M.; Bonizzoni, L.; Bruni, S.; Interlenghi, M.; Longoni, M.; Passaretti, A.; Caccia, M.; Salvatore, C.; Castiglioni, I.; et al. Imaging and Spectroscopic Data Combined to Disclose the Painting Techniques and Materials in the Fifteenth Century Leonardo Atelier in Milan. *Dye. Pigment.* **2021**, *187*, 109112. [CrossRef]

22. Alfeld, M.; Janssens, K.; Dik, J.; de Nolf, W.; van der Snickt, G. Optimization of Mobile Scanning Macro-XRF Systems for the in Situ Investigation of Historical Paintings. *J. Anal. At. Spectrom.* **2011**, *26*, 899–909. [CrossRef]
23. Alfeld, M.; Janssens, K. Strategies for Processing Mega-Pixel X-Ray Fluorescence Hyperspectral Data: A Case Study on a Version of Caravaggio's Painting Supper at Emmaus. *J. Anal. At. Spectrom.* **2015**, *30*, 777–789. [CrossRef]
24. Romano, F.P.; Caliri, C.; Nicotra, P.; Di Martino, S.; Pappalardo, L.; Rizzo, F.; Santos, H.C. Real-Time Elemental Imaging of Large Dimension Paintings with a Novel Mobile Macro X-Ray Fluorescence (MA-XRF) Scanning Technique. *J. Anal. At. Spectrom.* **2017**, *32*, 773–781. [CrossRef]
25. Alfeld, M.; Nolf, W.D.; Cagno, S.; Appel, K.; Siddons, D.P.; Kuczewski, A.; Janssens, K.; Dik, J.; Trentelman, K.; Walton, M.; et al Revealing Hidden Paint Layers in Oil Paintings by Means of Scanning Macro-XRF: A Mock-up Study Based on Rembrandt's "An Old Man in Military Costume". *J. Anal. At. Spectrom.* **2012**, *28*, 40–51. [CrossRef]
26. Kogou, S.; Lee, L.; Shahtahmassebi, G.; Liang, H. A New Approach to the Interpretation of XRF Spectral Imaging Data Using Neural Networks. *X Ray Spectrom.* **2020**. [CrossRef]
27. Mantler, M.; Schreiner, M.; Weber, F.; Ebner, R.; Mairinger, F. An X-Ray Spectrometer for Pixel Analysis of Art Objects. *Adv. X Ray Anal.* **1991**, *35*, 987–993. [CrossRef]
28. Alfeld, M.; de Viguerie, L. Recent Developments in Spectroscopic Imaging Techniques for Historical Paintings—A Review. *Spectrochim. Acta Part B At. Spectrosc.* **2017**, *136*, 81–105. [CrossRef]
29. Alberti, R.; Frizzi, T.; Bombelli, L.; Gironda, M.; Aresi, N.; Rosi, F.; Miliani, C.; Tranquilli, G.; Talarico, F.; Cartechini, L. CRONO: A Fast and Reconfigurable Macro X-Ray Fluorescence Scanner for in-Situ Investigations of Polychrome Surfaces. *X Ray Spectrom.* **2017**, *46*, 297–302. [CrossRef]
30. Ravaud, E.; Pichon, L.; Laval, E.; Gonzalez, V.; Eveno, M.; Calligaro, T. Development of a Versatile XRF Scanner for the Elemental Imaging of Paintworks. *Appl. Phys. A* **2015**, *122*, 17. [CrossRef]
31. Pouyet, E.; Barbi, N.; Chopp, H.; Healy, O.; Katsaggelos, A.; Moak, S.; Mott, R.; Vermeulen, M.; Walton, M. Development of a Highly Mobile and Versatile Large MA-XRF Scanner for in Situ Analyses of Painted Work of Arts. *X Ray Spectrom.* **2020**. [CrossRef]
32. Van der Snickt, G.; Dubois, H.; Sanyova, J.; Legrand, S.; Coudray, A.; Glaude, C.; Postec, M.; Van Espen, P.; Janssens, K. Large-Area Elemental Imaging Reveals Van Eyck's Original Paint Layers on the Ghent Altarpiece (1432), Rescoping Its Conservation Treatment. *Angew. Chem.* **2017**, *129*, 4875–4879. [CrossRef]
33. Mazzinghi, A.; Ruberto, C.; Castelli, L.; Ricciardi, P.; Czelusniak, C.; Giuntini, L.; Mandò, P.A.; Manetti, M.; Palla, L.; Taccetti, F. The Importance of Being Little: MA-XRF on Manuscripts on a Venetian Island. *X Ray Spectrom.* **2020**. [CrossRef]
34. Kogou, S.; Lucian, A.; Bellesia, S.; Burgio, L.; Bailey, K.; Brooks, C.; Liang, H. A Holistic Multimodal Approach to the Non-Invasive Analysis of Watercolour Paintings. *Appl. Phys. A* **2015**, *121*, 999–1014. [CrossRef]
35. Ricciardi, P.; Legrand, S.; Bertolotti, G.; Janssens, K. Macro X-Ray Fluorescence (MA-XRF) Scanning of Illuminated Manuscript Fragments: Potentialities and Challenges. *Microchem. J.* **2016**, *124*, 785–791. [CrossRef]
36. Duivenvoorden, J.R.; Käyhkö, A.; Kwakkel, E.; Dik, J. Hidden Library: Visualizing Fragments of Medieval Manuscripts in Early-Modern Bookbindings with Mobile Macro-XRF Scanner. *Herit. Sci.* **2017**, *5*, 6. [CrossRef]
37. Alfeld, M.; Baraldi, C.; Gamberini, M.C.; Walter, P. Investigation of the Pigment Use in the Tomb of the Reliefs and Other Tombs in the Etruscan Banditaccia Necropolis. *X Ray Spectrom.* **2019**, *48*, 262–273. [CrossRef]
38. Kozachuk, M.S.; Sham, T.-K.; Martin, R.R.; Nelson, A.J.; Coulthard, I.; McElhone, J.P. Recovery of Degraded-Beyond-Recognition 19th Century Daguerreotypes with Rapid High Dynamic Range Elemental X-Ray Fluorescence Imaging of Mercury L Emission. *Sci. Rep.* **2018**, *8*, 9565. [CrossRef] [PubMed]
39. Bonizzoni, L.; Maloni, A.; Milazzo, M. Evaluation of Effects of Irregular Shape on Quantitative XRF Analysis of Metal Objects. *X Ray Spectrom.* **2006**, *35*, 390–399. [CrossRef]
40. Brunetti, A.; Golosio, B. A New Monte Carlo Code for Simulation of the Effect of Irregular Surfaces on X-Ray Spectra. *Spectrochim. Acta Part B At. Spectrosc.* **2014**, *94–95*, 58–62. [CrossRef]
41. Trojek, T. Reduction of Surface Effects and Relief Reconstruction in X-Ray Fluorescence Microanalysis of Metallic Objects. *J. Anal. At. Spectrom.* **2011**, *26*, 1253–1257. [CrossRef]
42. Saleh, M.; Bonizzoni, L.; Orsilli, J.; Samela, S.; Gargano, M.; Gallo, S.; Galli, A. Application of Statistical Analyses for Lapis Lazuli Stone Provenance Determination by XRL and XRF. *Microchem. J.* **2020**, *154*, 104655. [CrossRef]
43. Weyermann, J.; Schläpfer, D.; Hueni, A.; Kneubühler, M.; Schaepman, M. *Spectral Angle Mapper (SAM) for Anisotropy Class Indexing in Imaging Spectrometry Data*; Shen, S.S., Lewis, P.E., Eds.; International Society for Optics and Photonics: San Diego, CA, USA, 2009; p. 74570B.
44. Panchuk, V.; Yaroshenko, I.; Legin, A.; Semenov, V.; Kirsanov, D. Application of Chemometric Methods to XRF-Data—A Tutorial Review. *Anal. Chim. Acta* **2018**, *1040*, 19–32. [CrossRef] [PubMed]
45. Leardi, R. (Ed.) *Nature-Inspired Methods in Chemometrics: Genetic Algorithms and Artificial Neural Networks*, Data Handling in Science and Technology, 1st ed.; Elsevier Science: Amsterdam, The Netherlands, 2003; Volume 23, ISBN 978-0-444-51350-2.
46. Galli, A.; Caccia, M.; Alberti, R.; Bonizzoni, L.; Aresi, N.; Frizzi, T.; Bombelli, L.; Gironda, M.; Martini, M. Discovering the Material Palette of the Artist: A p-XRF Stratigraphic Study of the Giotto Panel 'God the Father with Angels': Discovering the Pigment Palette Using a p-XRF Stratigraphic Analysis. *X Ray Spectrom.* **2017**, *46*, 435–441. [CrossRef]

47. Ciatti, M.; Seidel, M. (Eds.) *Giotto. La Croce di Santa Maria Novella*; Problemi di Conservazione e Restauro; EDIFIR: Firenze, Italy, 2003; ISBN 88-7970-106-1.
48. Romano, S.; Petraroia, P. *Giotto, l'Italia. Catalogo della Mostra*; Illustrated Edizione; Mondadori Electa: Milano, Italy, 2015; ISBN 978-88-918-0513-3.
49. Bonizzoni, L.; Galli, A.; Poldi, G. In Situ EDXRF Analyses on Renaissance Plaquettes and Indoor Bronzes Patina Problems and Provenance Clues. *X Ray Spectrom.* **2008**, *37*, 388–394. [CrossRef]
50. Barcellos Lins, S.A.; Ridolfi, S.; Gigante, G.E.; Cesareo, R.; Albini, M.; Riccucci, C.; di Carlo, G.; Fabbri, A.; Branchini, P.; Tortora, L. Differential X-Ray Attenuation in MA-XRF Analysis for a Non-Invasive Determination of Gilding Thickness. *Front. Chem.* **2020**, *8*, 175. [CrossRef]
51. Saverwyns, S.; Currie, C.; Lamas-Delgado, E. Macro X-Ray Fluorescence Scanning (MA-XRF) as Tool in the Authentication of Paintings. *Microchem. J.* **2018**, *137*, 139–147. [CrossRef]
52. Schowengerdt, R. *Remote Sensing; Models and Methods for Image Processing*, 3rd ed.; Elsevier: Amsterdam, The Netherlands, 2006; ISBN 978-0-12-369407-2.
53. Daniel, F.; Mounier, A.; Pérez-Arantegui, J.; Pardos, C.; Prieto-Taboada, N.; de Vallejuelo, S.F.O.; Castro, K. Hyperspectral Imaging Applied to the Analysis of Goya Paintings in the Museum of Zaragoza (Spain). *Microchem. J.* **2016**, *126*, 113–120. [CrossRef]
54. Kruse, F.A.; Richardson, L.L.; Ambrosia, V.G. Techniques Developed for Geologic Analysis of Hyperspectral Data Applied to Near-Shore Hyperspectral Ocean Data. In Proceedings of the ERIM 4th International Conference, Remote Sensing for Marine and Coastal Environments: Environmental Research Institute of Michigan (ERIM), Ann Arbor, MI, USA, 17–19 March 1997; Volume I, pp. I-233–I-246.

Article

Synchrotron X-ray Microprobes: An Application on Ancient Ceramics

Alessandra Gianoncelli [1,*], **George Kourousias** [1], **Sebastian Schöder** [2], **Antonella Santostefano** [3], **Maëva L'Héronde** [4], **Germana Barone** [5], **Paolo Mazzoleni** [5] and **Simona Raneri** [6]

[1] Elettra—Sincrotrone Trieste, Strada Statale 14, km 163.5 in Area Science Park, 34149 Basovizza, Italy; george.kourousias@elettra.eu

[2] Synchrotron SOLEIL, PUMA Beamline, Saint-Aubin BP48, 91192 Gif-sur-Yvette, France; sebastian.schoeder@synchrotron-soleil.fr

[3] Department of Ancient and Modern Civilizations, University of Messina, Polo Universitario dell'Annunziata, 98168 Messina, Italy; antonellasantostefano@yahoo.it

[4] IPANEMA USR3461, CNRS, Université Paris-Saclay, Ministère de la Culture, UVSQ, MNHN, 91192 Saint-Aubin, France; maeva.lheronde@synchrotron-soleil.fr

[5] Department of Biological, Geological and Environmental Sciences, University of Catania, Corso Italia 57, 95129 Catania, Italy; gbarone@unict.it (G.B.); pmazzol@unict.it (P.M.)

[6] Institute of Chemistry of Organometallic Compounds, National Research Council, ICCOM-CNR, Via G. Moruzzi 1, 56124 Pisa, Italy; simona.raneri@pi.iccom.cnr.it

* Correspondence: alessandra.gianoncelli@elettra.eu

Abstract: Synchrotron X-ray μ- and nano-probes are increasingly affirming their relevance in cultural heritage applications, especially in material characterization of tiny and complex micro-samples which are typical from archaeological and artistic artifacts. For such purposes, synchrotron radiation facilities are tailoring and optimizing beamlines and set-ups for CH, taking also advantages from the challenges offered by the third-generation radiation sources. In ancient ceramics studies, relevant information for the identification of production centers and manufacture technology can be obtained in a non-invasive and non-destructive way at the micro-sample level by combining different SR based methods. However, the selection of appropriate beamlines, techniques and set-ups are critical for the success of the experiments. Fine and varnished wares (e.g., Attic and western-Greek colonial products) are an excellent case study for exploring challenges offered by synchrotron X-ray microprobes optimized to collect microchemical and phase-distribution maps. The determination of provenance and/or technological tracers is relevant in correctly classifying productions, often based only on ceramic paste, gloss macroscopic features or style. In addition, when these vessels are preserved in Museums as masterpieces or intact pieces the application of non-invasive approach at the micro sample is strictly required. Well-designed synchrotron μXRF and μXANES mapping experiments are able providing relevant clues for discriminating workshops and exploring technological aspects, which are fundamental in answering the current archaeological questions on varnished Greek or western-Greek colonial products.

Keywords: X-ray fluorescence; synchrotron radiation; μXRF; μXANES; black gloss; ancient ceramics

Citation: Gianoncelli, A.; Kourousias, G.; Schoeder, S.; Santostefano, A.; L'Héronde, M.; Barone, G.; Mazzoleni, P.; Raneri, S. Synchrotron X-ray Microprobes: An Application on Ancient Ceramics. *Appl. Sci.* **2021**, *11*, 8052. https://doi.org/10.3390/app11178052

Academic Editors: Letizia Bonizzoni and Anna Galli

Received: 23 July 2021
Accepted: 26 August 2021
Published: 30 August 2021

Publisher's Note: MDPI stays neutral with regard to jurisdictional claims in published maps and institutional affiliations.

1. Introduction

X-ray fluorescence (XRF) for cultural heritage materials is a widely used and well-assessed technique for compositional characterization of archaeological and artistic objects. It enables the elemental analysis of materials and provides an easy way to determine the materiality of artifacts [1–4].

In ancient ceramics studies, laboratory XRF is traditionally used for provenance issues. XRF is a non-invasive technique, this means that it can be used directly on the object without the need of sampling. However, when used with a non-focused X-ray beam, as it is often provided by laboratory instruments, sampling might be needed to obtain meaningful and reliable bulk analysis. This typically consists of the preparation of pressed pellets from a few grams of powdered sample. In addition, it is sometimes combined with other destructive analytical tools for trace elements determination, such as ICP-MS, ICP-OES, NAA [5]. For provenance studies, geochemical data are often processed by using statistical methods able to create correlation among group of samples, also in comparison with databases [6,7]. When geochemical tracers fail in group classification and provenance discrimination, recent studies have demonstrated the merits of isotopic analysis as clay provenance fingerprint [8,9]. Ceramics are the most numerous records in archaeological excavations and are often expendable for destructive analysis. However, in some cases, the artistic and cultural value of ceramic objects—especially when preserved and exhibited in Museums as masterpieces—prevents macro-sampling needed for such destructive XRF analyses. The use of portable XRF systems enables to overcome this limit Being nondestructive and noninvasive they offer the advantage of material characterization without sampling and directly in situ [10–13]. Indeed, the advent of new powerful and focused X-ray tubes in the last decades, together with performant detectors, which do not require liquid nitrogen cooling, has pushed to use and the performance of both portable and laboratory XRF systems. Single point XRF equipment is currently available in the majority of diagnostic laboratories. Moreover, the accessibility of advanced macro-XRF systems introduced in laboratories—pioneering set-ups designed at synchrotron radiation facilities—allows determining and localizing the distribution of chemical elements at the sample surface [14–16]. Detection limits and element ranges are the most common limits of portable and lab-based systems. Thus, when micro-sampling is allowed (for example, from hidden part of the ceramic vessel) the use of synchrotron-based (SR) X-ray sources appears quite valuable. It is non-invasive and non-destructive on the micro-samples, which can be later on used for further investigations. SR X-ray sources offer numerous advantages. They are brilliant sources assuring intense X-ray radiation, providing a quasi-monochromatic beam and an energy selection over a wide range, which can be chosen according to elements of interest [17,18].

In the last decades, one of the main innovations at SR-based X-ray sources are micro- and nano-focusing systems aimed at obtaining micro- and nanometric lateral resolution for both qualitative and quantitative analysis in complex and/or tiny samples [19–21]. Soft and hard X-ray microprobes are available at synchrotron radiation sources. The selection of the probes depends on the energy range of interest—and thus the elements to detect. In ceramic studies, elements with emission energies below 10 keV are common constituents. However, the use of harder X-ray microprobes appears suitable for the determination of elements usually present in traces and also relevant in provenance or technological studies. Among SR-based X-ray microbeam techniques, μXRF mapping systems appear particularly useful in analyzing ceramic decorations (slip, glazes, etc.) or to detect enrichment/depletion of specific elements across ceramic section for provenance or technological purposes [22,23]. Being heterogenous materials, the analysis of ceramics can benefit from the combination of different SR-based X-ray methods with micrometric or sub-micrometric spatial resolutions. For example, μXRF can be combined with micro-X-ray diffraction (μXRD) for the determination of crystalline phases in both the ceramic body and decorative layers or nano-XRD for single crystal studies even in not-crystalline matrix, such as in glazes [24–26]. μXRF can be also coupled with μ- or nano-XANES, enabling the determination and even the distribution of the valence state of an element [27]. Usually, this method requires the preliminary acquisition of a μXRF map for the selection of areas of interest. The XANES measurement consists of scans in fluorescence or transmission (TXM) mode—depending on sample preparation and characteristics—at defined energies across the adsorption edge of the element of interest. μ- and nano-XANES are still relatively recent application in the CH field and therefore poorly explored [20]. In literature, very few examples discuss the merits of μXANES in the mild or hard X-ray range, particularly suitable for technological studies on slips and gloss layers in decorated vessels [28–30].

The combination of SR-based X-ray methods, such as fluorescence, diffraction and absorption, appears relevant for different purposes in ancient ceramics studies. For provenance issues, SR-based μXRF mapping might provide insights on the chemical distribution of elements across the sample section. It can highlight enrichment/depletion of minor and trace elements in ceramic paste vs. surface decoration, also in comparison with other productions and/or reference data. Additionally, SR-based X-ray methods are suited for technological studies. The microanalysis of slip or varnished surfaces might provide information on chemical composition, elemental distribution, presence of crystallites in glossy matrix and speciation of elements (metals) for the better understanding of manufacture procedures.

For such applications, the sampling of small fragments or even a more refined sample preparation (e.g., cross-section or thin section) depends on the field of view required in the investigation. XRF or XANES maps acquired with μbeams in fluorescence mode can be carried out on small fragments. The sample can be positioned in cross-sectional geometry to acquire information on both the clay paste and the surface slip. Otherwise, the use of nano-beams for acquiring details on single layers, characterize crystallites, investigating the element distribution at sub-micrometric scale, or performing transmission XANES maps require the preparation of thin sections or microtome sections. The use of μ and nano-beams implies in general higher localized radiation doses in the sample, as the flux is focused on a smaller spot size. However, in ceramic studies radiation damage is usually not a big issue. Sample preparation depends also on beamline set-up; for example, it is possible to perform the analysis in air—even on big objects or entire vessels—or in vacuum—which often requires microtome sections. Looking at the International Synchrotron Radiation Facilities, Table 1 reports a list of beamlines particularly suitable for material sciences applications, some of which are specifically optimized for cultural heritage studies, with details about energy range (which elements can be detected, mapped and for which the speciation can be determined), beam size (level of resolution) and sample environment (sample preparation required, vacuum or air, big or small samples). The overall listed information appears quite relevant in designing a successfully experiment, depending on the archaeological question to be answered.

Among ancient Greek ceramics, red and black figures wares and black-gloss wares have a great interest both from the stylistic/typological point of view and the provenance and technological aspects. In fact, their analysis might draw the mobility of goods and/or painters and artisans through Greece and the western-Greek colonies. The identification of specific chemical markers to discriminate productions—especially among western-Greek colonies—is not straightforward due to high depuration of raw materials. In addition, even if the current literature mainly agrees on the technological routine applied in ancient time to obtain well manufactured red, black or red and black gloss wares, some issues remain still open [29,31,32]. To explore the challenges offered by SR-based X-ray microbeams in provenance and technological studies of ancient ceramics, fragments of black-gloss ware from different Greek and western Greek colonial products have been selected—as examples—and investigated at PUMA beamline of the SOLEIL Synchrotron Radiation Facility [33].

The selection of this beamline allowed (i) the determination of elements with emission energies higher than 10 keV, relevant in provenance and technological studies; (ii) obtaining micrometric lateral resolution (microbeam) for the determination of enrichment/depletion of elements in layers micrometric in thickness (black gloss); (iii) collecting both XRF and XANES maps at micrometric scale; (iv) to use samples without specific sample preparation due to the analysis being carried out in air, with a relatively flexible sample stage to accommodate fragments with different shapes or even entire vessels.

For the purpose of comparison, other classical non-destructive and non-invasive (at micro-sample level) microchemical methods were used and the results are discussed in the light of the SR X-ray data.

Table 1. List of synchrotron beamlines suitable for materials science applications (especially cultural heritage) with the indication of available techniques and some practical and useful set-up information.

Beamline	Facility	Country	Energy Range [keV]	Available Techniques	Beam Size [um]	Sample Environment	Applications
PUMA	SOLEIL	France	7–22	μXRF, μXANES, μXRD	3.5 × 3.5	Air	CH (70%), Environmental sciences (30%)
LUCIA	SOLEIL	France	0.8–8	μXRF, μXANES	2 × 3	Vacuum	Life Sciences, Materials sciences, CH
ID21	ESRF	France	2–11	μXRF, μXANES, μXRD	0.03 × 0.07	Vacuum	Life Sciences, Materials sciences, CH
ID16B	ESRF	France	6–65	μXRF, μXANES, μXRD	0.05 × 0.05	Air	Life Sciences, Materials sciences, CH
TwinMic	Elettra	Italy	0.2–2.2	μXRF, μXANES, STXM	circular, diameter from 0.1 to 2.5	Vacuum	Life Sciences, Materials sciences, CH
XFM	Australian Synchrotron	Australia	4.1–27	μXRF, μXANES	circular, diameter from 1 to 5	Air	CH, Life Sciences, Materials Science
I08	Diamond	UK	0.2–4.2	μXRF, μXANES, STXM	circular, diameter from 0.1 to 2	Vacuum	Life Sciences, Materials sciences, CH
NanoMAX	MAXIV	Sweden	6–28	μXRF, μXANES, ptychography	0.05 to 0.2	Air	Life Sciences, Materials sciences, CH
SoftiMAX	MAXIV	Sweden	0.275–2.5	μXRF, μXANES, STXM	0.01 to 0.1	Vacuum	Life Sciences, Materials sciences, CH
26-ID	APS	USA	6–12	μXRF, μXRD	0.03 × 0.03	Air	Life Sciences, Materials sciences, CH
20-BM-B	APS	USA	2.7–32.7	μXRF, μXANES	5 × 5 or 25 × 25	Air	Life Sciences, Materials sciences, CH
20-ID-B,C	APS	USA	4.3–27 or 8–50	μXRF, μXANES	2 × 2	Air	Life Sciences, Materials sciences, CH
SM	CLS	Canada	0.13–2.7	μXRF, μXANES, STXM	0.03	Vacuum	Life Sciences, Materials sciences, CH
SGM	CLS	Canada	0.25–2	μXRF, μXANES	1 × 0.1	Vacuum	Life Sciences, Materials sciences, CH
VESPER	CLS	Canada	6–30	μXRF, μXANES, μXRD	from 2 to 4	Air	Life Sciences, Materials sciences, CH
HXN	NSLS II	USA	12–17	μXRF, μXANES, μXRD, 3D XRF	from 0.1 to 0.4	Air	Life Sciences, Materials sciences, CH
XFM	NSLS II	USA	4–20	μXRF, μXANES	from 1 to 10	Air	Life Sciences, Materials sciences, CH

2. Materials and Methods

In this case, 14 samples of black-gloss ceramics from the excavations of the Greek colonies of Gela and Messina (Sicily, Italy) have been selected for this study. They are representative of two different productions, Laconian (GEL 1–8) and so-called Chalciadian (ME 67–70, 73, 133). The classification was preliminarily based on the morphological and typological features of the selected specimens and on macroscopic observations of the black-gloss and the paste (Table 2). Attic products and Sicilian (Geloan) and south-Italian (Locrian?) colonial products already characterized in a previous research [22] were also re-considered for the purpose of comparison.

The so-called Chalcidian pottery is known in literature for both figurative and aniconic series, dated from the 6th century B.C. to the beginning of the 5th century B.C. According to the most recent archaeological studies, it is a colonial product that has been located in the southern Calabria or in the Strait of Messina area [34–37]. Laconian black pottery is produced in the Greek region of Laconia during the 6th century B.C. and is exported to different sites in the Mediterranean [38,39].

For the experiments, small fragments were sampled from vessels preserved in Archeological Museums of Gela and at the deposits of the Superintendence of Messina. Vessels from Messina classified as Chalcidian were also studied by benchtop XRF and ICP-MS in the frame of previous published research, pointing out some geochemical criteria for productions discrimination, which are based—however—on destructive methods [35].

Non-destructive and non-invasive preliminary studies [22] at micro-samples on Attic, Sicilian and South-Italian colonial products pointed out enrichment-depletion of specific elements in clay paste and black gloss useful as provenance indicators. Additionally, Zn has been verified as peculiar of non-Sicilian colonial products and of specific manufacturing practices [22], of which nature and workshop distribution needs to be verified.

The analysis of the selected corpus by different non-destructive and non-invasive methods offered the opportunity to explore the challenges offered by the combination of SR based μXRF and μXANES mapping in providing valuable provenance and technical features for this ceramic class.

A very preliminary qualitative chemical characterization was obtained by portable XRF mapping using an Elio Bruker device equipped with an x-y motor stage and an X-ray tube with Rhodium anode. The measuring spot on the surface was about 1 mm. 2D maps were acquired on the gloss surface using a 40kV tension of the X-ray tube, 80 μA current, 3 s/point acquisition time. However, the large analysis spot (about 1 mm) prevented the necessary resolution for accurate determination of the black-gloss features; in addition, the analysis is expected to be influenced by bulk composition.

Following a successful microchemical approach on black-gloss potteries [22] micro-samples were analysed at PUMA beamline, Synchrotron Soleil (France). A KB mirror focuses the X-ray photons to a spot of 3 μm $\times$ 3 μm on the sample. The surface to be analyzed is orientated at an angle of 45° in respect to the beam axis, producing an effective beam size of 3 μm in vertical and 4.3 μm in horizontal direction. The measurements were performed in ambient air and temperature. The XRF signal was acquired by a SGX Sirius SD silicon drift detector installed at 90° from the incident beam. A visible light microscope located perpendicularly to the sample surface allowed micrometric visualization and navigation of the sample. During XRF analyses, the samples were scanned at 18 keV with a step size of 10 μm using 1 s acquisition time per pixel. Samples were scanned in two geometries, namely with the gloss facing the beam and in cross-section. On these latter samples, after XRF mapping, specific points on the surface of the gloss were selected to perform XANES measurements across the Fe K and Zn K absorption edges. The XANES spectra were collected in fluorescence mode with 1 s/point acquisition time. For Fe an energy resolution of 2 eV was chosen in the pre-edge range from 7.03 to 7.08 keV and in the post-edge zone from 7.2 to 7.33 keV. The area around the edge from 7.08 to 7.2 keV was scanned with higher resolution in steps of 0.5 eV. For Zn an energy resolution of 2 eV was used in the pre-edge range from 9.37 to 9.4 keV and in the post-edge range from 9.47 to 9.71 keV, while close to the edge from 9.4 to 9.47 KeV an energy resolution of 0.5 eV/step.

Reference standards of Fe^{2+} and Fe^{3+} were used for calibration and to help Fe-speciation identification, while ZnO and $ZnFe_2O_4$ (Gahnite) were used for calibration and help with Zn-speciation. Collected spectra were compared both to the measured reference standards and to reference spectra available in the literature.

For a pre-selection of the samples to be mapped in cross-section, XANES maps were acquired at 4 different energies across Fe and Zn K absorption edges, 7330 eV, 7136 eV, 7132 eV, 7127 eV for Fe and 9872 eV, 9688 eV, 9669 eV and 9665 eV for Zn, respectively. Some of these energies were used to acquire insights on Fe (7127 eV, 7132 eV) and Zn (9688 eV, 9669 eV and 9665 eV) speciation distribution in 2D, in addition to the single point spectra previously acquired. XANES maps were acquired over areas of 150 μm $\times$ 150 μm, with 5 μm step size and 1 s acquisition time per point, centering the area as 50 μm above surface and 100 μm below surface (that is, inside the sample paste).

Table 2. List of studied samples.

Sample ID	Type and Chronology	Provenance	Attribution (Based on Typology and Black Gloss Appearance)
GEVN 1–6, GEVN 8, GEVN 11–13 [22]	Cup-skyphos (GEVN 1), skyphoi (GEVN 4, 8, 12–13), kylikes (GEVN 3, 5–6), small bowls (GEVN 2, 11) (5th-first quarter 4th cent. B.C.)	Gela, Molino a Vento, excavation 1955–1956	Sicilian colonial production (Geloan)
GEVN 14–15, 17–20 [22]	Skyphoi (GEVN 14–15), kylikes (GEVN 18, 20), small bowls (GEVN17, 19) (5th-first quarter 4th cent. B.C.)	Gela, Old Station, excavation 1984	Sicilian colonial production (Geloan)
GEVN 7 [22]	Cup-skyphos (last quarter 6th cent. B.C.)	Gela, Molino a Vento, excavation 1956	Attic production
GEVN 9 [22]	Saltcellar (Ca. 450 B.C.)	Gela, Molino a Vento, excavation 1956	Attic production
GEVN 10 [22]	Stemmed dish (late 6th-early 5th cent. B.C.)	Gela, Molino a Vento, excavation 1955	Chalcidian production
GEVN 16 [22,40,41]	Skyphos (Ca. 400–380 B.C.)	Gela, Old Station, excavation 1984	South-Italian production (Locrian?)
GEL 1–8 [22]	Kraters (6th cent. B.C.)	Gela, Molino a Vento, excavation 1955–1956, 1974	Laconian production
ME 67 (Inv. 10561), ME 68 (Inv. 10566), ME 69 (Inv. 10564), ME 70 (Inv 10565), ME 73 (Inv. 10563), ME 133 (Inv. 10544) [35,37]	Skyphoi (ME 67–69, 73, 133), krater (ME 70) (second half 6th–early 5th cent. B.C.)	Messina, Via Industriale-Isolato S, US 3. Excavation 1991–1992	Chalcidian production

For comparison, on the same analysis area scanned at PUMA, traditional microchemical investigation was performed by SEM-EDS at IPANEMA facility labs [42] by using a ZEISS Supra55VP SEM-EDS with a Schottky Field Emission Gun (FEG) equipped with a Bruker EDS system. Measurements were carried out without any metal coating on the sample surface to assure a non-invasive and non-destructive testing; nevertheless, to obtain better images and microanalysis, studied samples were wrapped in aluminum foil leaving out a small window in the region of interest to analyze.

3. Results

3.1. Portable XRF

Even with obvious limits, a fast and non-destructive scan of the surface provided some clues on compositional differences among the studied productions. In fact, Attic and Laconian vessels, and Chalcidian, south-Italian (Locrian?) and Sicilian colonial (Geloan) samples show different features.

In Attic fragments glaze is mainly Fe-rich with low amounts of silicon and potassium. Laconian samples are characterized by a Fe-based black-gloss with Mn co-localization, along with silicon and potassium in lower amount. Sicilian colonial fragments are characterized by Fe-based gloss with potassium and manganese. On the other hand, Chalcidian and Locrian products show a co-localization of Fe and Zn, along with silica and low amount of K and Mn. Of course, this qualitative analysis provided a not-univocal classification criteria. However, a portable XRF approach seems to permit the discrimination between Attic and colonial products, and between Sicilian and South-Italian colonial products based on Zn marker (Figure S1).

3.2. SR Based X-ray Methods Using Microprobes

3.2.1. μXRF Maps

μXRF maps collected on representative samples of black gloss ceramics allowed to distinguish different productions. It also showed the usefulness of the cross-sectional set-up in data collection to detect enrichment/depletion in gloss vs. clay paste (Figure 1). The different productions examined were thus non-destructively discriminated based on microchemical tracers identified on the black gloss vs. the clay paste.

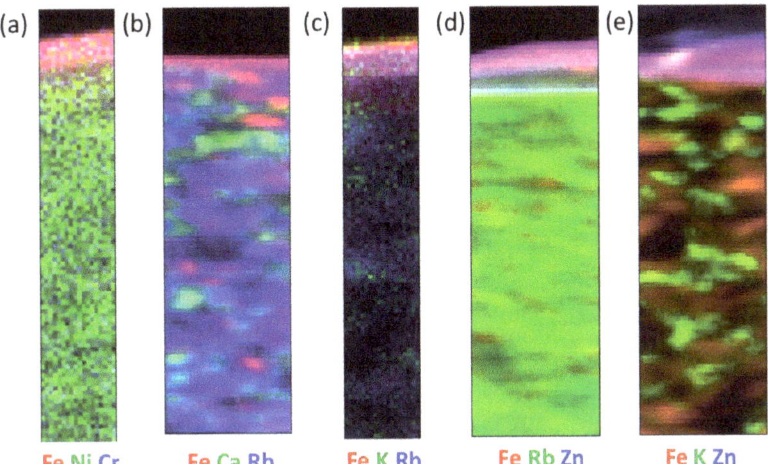

Figure 1. RGB correlation considering relevant elemental markers in cross-section geometry for (**a**) Attic (GEVN9, 100 × 500 μm), (**b**) Laconian (GEL5, 150 × 500 μm), (**c**) Sicilian (GEVN15, 100 × 500 μm), (**d**) Chalcidian (ME68, 150 × 500 μm) and (**e**) Locrian (?) (GEVN10, 150 × 500 μm) products.

The thin (~20 μm) shiny, compact, deep black gloss proper of Attic production is characterized by a prominent Fe-based black-gloss, and a clay paste enriched in Ni and Cr. Laconian vessels are characterized by a very thin (about 10 μm) matte black gloss characterized by Fe-rich composition and a clay paste rich in Rb and Ca. The thinner (~15 μm), matte and brownish (even reddish) gloss typical of vessels from the Sicilian Greek colony of Gela is due to Fe-based black-gloss enriched in Rb and Mn, and a K-rich clay paste.

South Italian colonial vessels (Chalcidian and probable Locrian products) showed peculiar features. The matte and compact black/black-bluish gloss is in fact characterized by Fe- and Zn enrichment and a clay paste marked by Ni, Cr signature (less than in Attic products), while Rb, K, Mn in Chalcidian black gloss ceramics. Accordingly, previous mineralogical and geochemical investigations on reference groups assessed characteristics geochemical fingerprints [35,37] for this production, providing criteria based on destructive methods useful to locate the products in the south-Italian Greek colonies (Strait of Messina area and Ionian cost) and discriminate them from Sicilian and Attic ones. However, bulk analysis could not trace a quite interesting chemical marker, which is also useful for ceramic technological studies. The SR based non-invasive investigation revealed this to be a Fe-Zn-rich black gloss.

3.2.2. μXANES Spectra and Maps

XANES Fe K-edge spectra acquired on the black gloss layer—with the sample cross-section facing the incoming beam—show the presence of both Fe^{2+} and Fe^{3+} phases, as already found in some of these glosses analysed in our previous work [22]. As it can be seen in Figure 2, spectra collected on Laconian and Chalcidian samples exhibit in some cases (GEL2, GEL3, GEL5, GEVN10, ME68 and ME70) Fe^{3+} preferentially, while in others (GEL1, GEL8, GEVN7, GEVN16, ME67 and ME133) a mixture of both phases. This is visible both in the pre-edge (panels in Figure 2a,b) and edge analysis (Figure S2).

XANES Fe-K edge maps collected on ME68, ME70, GEVN7, GEVN, GEVN16, GEL2 and GEL5 samples in cross-section geometry add further information to the point XANES spectra. We collected XRF maps at different energies across Fe K-edge. After normalization, we could discriminate whether Fe^{3+} is the dominant phase by differentiating energies E1 and E2 in Figure S3. The processing is shown in Figure 3, where each differential map is depicted beside its corresponding Fe maps acquired at 18 keV in cross-section geometry. GEVN7 shows a sharp gloss layer where the ratios between the abovementioned energies implies a combination of both Fe^{2+} and Fe^{3+}, followed by GEVN16 where Fe^{2+} is present but in less proportion. It is followed by GEVN10 and GEL2, where the gloss layer is less defined (see corresponding Fe map) and finally ME68, where the ratio varies along the gloss layer indicating a slight inhomogeneity between the Fe^{2+} and Fe^{3+} distribution. In GEL2 and ME70 the ratio between the two energies is smaller, highlighting a more pronounced predominance of Fe^{3+} over Fe^{2+}. Overall, the Fe XANES maps confirm the single point XANES spectra but allow a better special visualization of the distribution of the ratio between the two phases.

Zn-edge XANES spectra were also collected in cross-section mode on a sub-set of the glosses exhibiting Zn on the gloss surface. All of them show a very similar spectrum (Figure 4), which can be mainly attributed to $ZnAl_2O_4$ (Figure S4). Point XANES spectra were collected also on other glosses (ME67, ME70 and ME133). Even if the spectra turned out to be noisy due to lower Zn content they also confirm the presence of mainly $ZnAl_2O_4$.

XANES Zn-K edge maps collected on GEVN10, GEVN16, ME68—as example of the so-called Chalcidian products (southern Calabria and the Strait of Messina area)—show a predominance of $ZnAl_2O_4$ along all the gloss surface layer, as also found in [43]. Figure 5 shows the Zn XRF maps in all three samples and the obtained distribution of $ZnAl_2O_4$ phase, shown in red in panels a, b and c, respectively. The last image was obtained by differential imaging evaluating the ratios between different peaks identified from reference standards spectra (Figure S4), as successfully used in [44]. In particular we evaluated the ratios between the peaks 3 to 2, 3 to 1 and 2 to 1, which turned out to be bigger than one for the first two ratios and very close to one for the last one, confirming $ZnAl_2O_4$ phase on the surface.

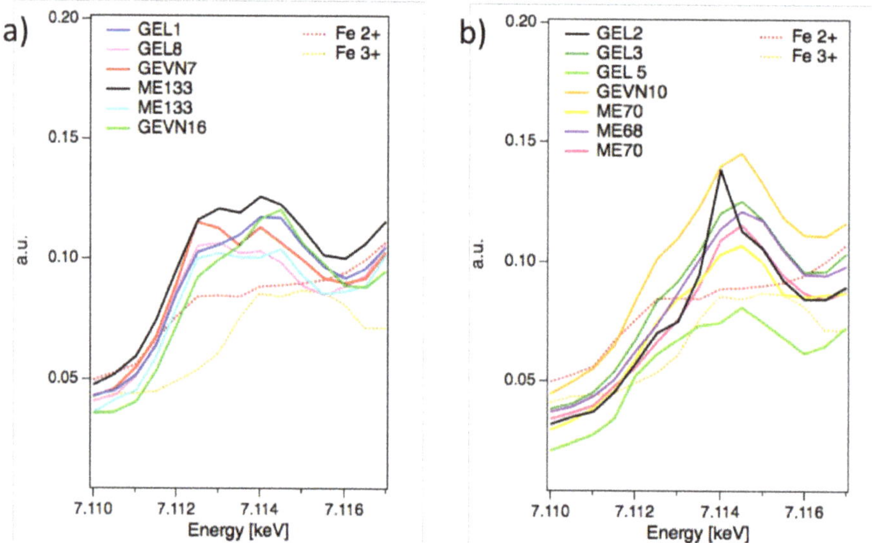

Figure 2. Fe-XANES spectra acquired on a 3 μm × 4.2 μm spot on the gloss layer of the samples indicated in the legend, all mounted in cross section geometry. Samples of panels (**a**) exhibit a mixture of both Fe^{2+} and Fe^{3+}, while samples on panel (**b**) mainly Fe^{3+}.

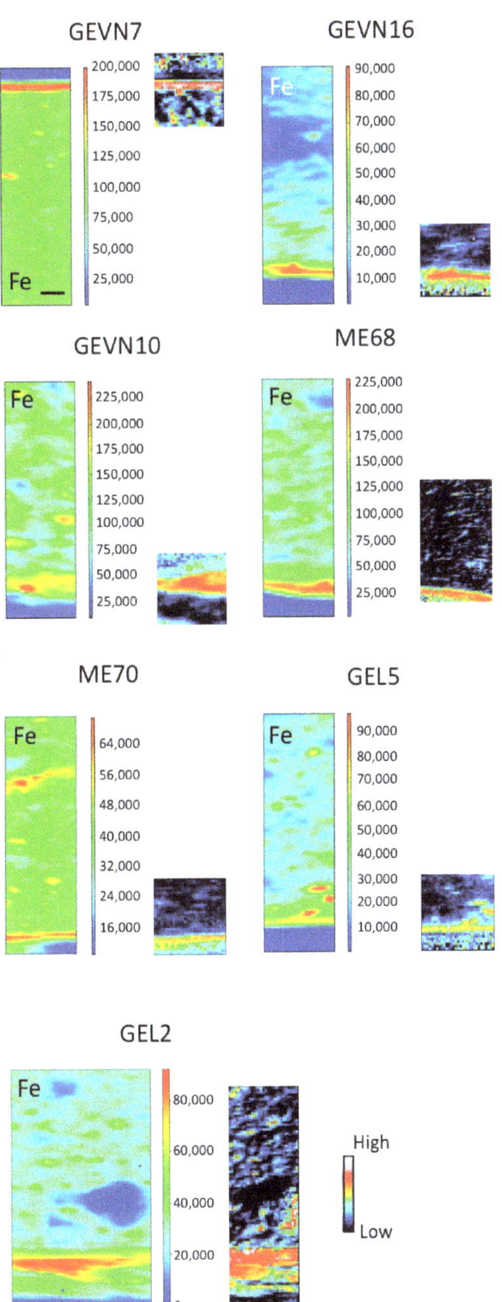

Figure 3. Fe XRF maps of sample GEVN7, GEVN10, GEVN16, ME68, ME70, GEL2 and GEL5 in cross section geometry, together with the corresponding distribution of the ratio between the maps collected at 7127 eV and 7132 eV. Scale bar is 50 μm and is valid for all images.

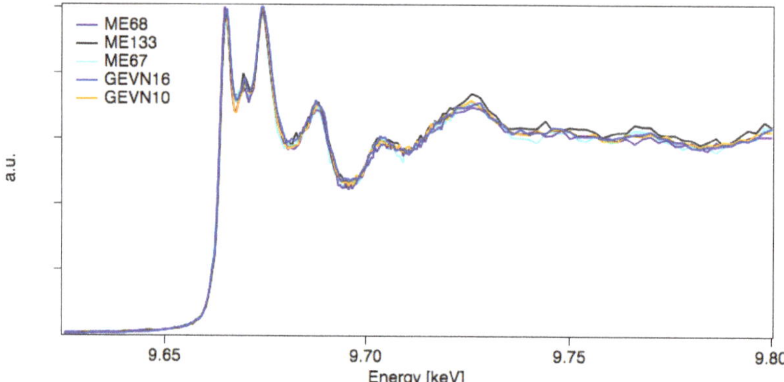

Figure 4. Zn XANES spectra acquired on a 3 μm × 4.2 μm spot on the gloss layer of samples ME68, ME133, ME67, GEVN10 and GEVN16 in cross section geometry.

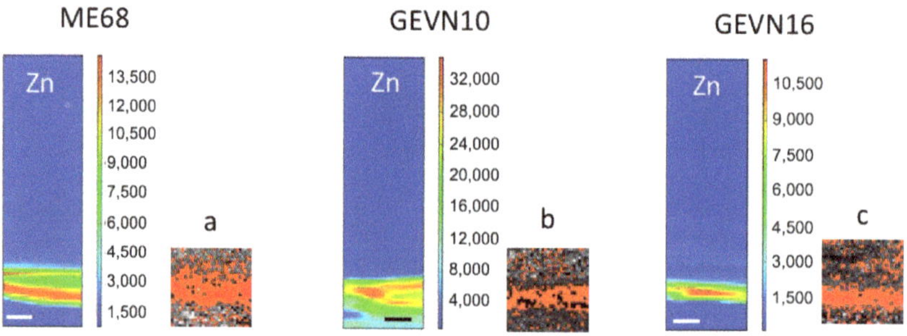

Figure 5. Zn XRF maps of sample ME68, GEVN10 and GEVN16 in cross section geometry, together with the corresponding distribution of ZnAl$_2$O$_4$ phase, depicted in red in panels (**a**–**c**). Scale bar is 50 μm.

3.3. Comparative SEM-EDS Analysis

Morphological and microchemical characterization of the slip has been determined by SEM-EDS measurements. SEM images and chemical maps show the distribution of Al, Si, K, Mg, Ti and Fe both on the slip and the clay paste (Figures 6–8). In all the inspected samples, the clay paste is highly vitrified, claiming for ~900–950 °C maximum firing temperature at the oxidizing conditions. The gloss is smooth and mainly composed by Si, Al, Fe and K. It is quite uniform in Attic production (Figure 6), showing a sharp separation from the body. In colonial South-Italian (southern Calabria and the Strait of Messina area; Figure 7) and Laconian (Figure 8) products the gloss layer fades out into the body, indicating a different technological routine during the first oxidizing stage. In the black-gloss, Mg and Ti are also present, possibly related to Fe-substitution. Looking at the silica, alumina and Fe contents, the Laconian products have lower tenors in these elements, which indicate the provenance of clays used for the black gloss (Figure S5). For the other products, no substantial differences can be appreciated, while groups are clearly discriminated by coupled μXRF and μXANES maps (Table 3).

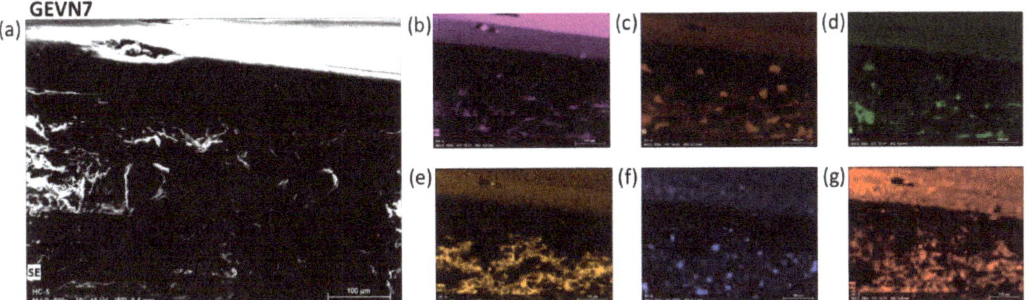

Figure 6. (**a**) SEM image of sample GEVN7; maps of (**b**) Al, (**c**) Si, (**d**) K, (**e**) Mg, (**f**) Ti and (**g**) Fe.

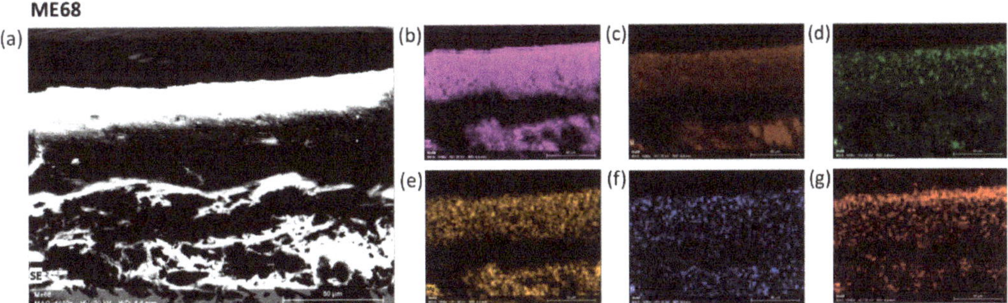

Figure 7. (**a**) SEM image of sample ME68; maps of (**b**) Al, (**c**) Si, (**d**) K, (**e**) Mg, (**f**) Ti and (**g**) Fe.

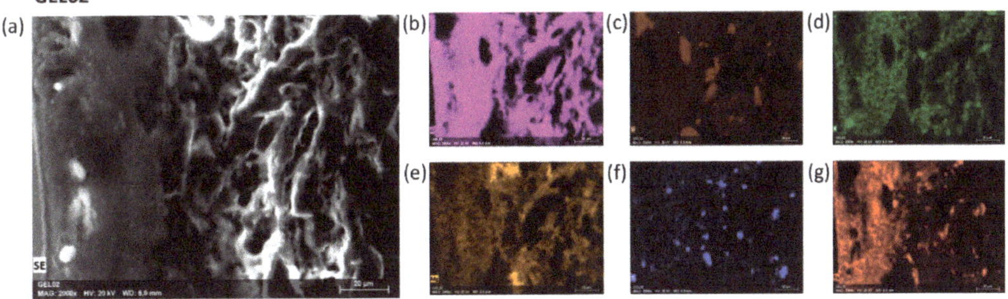

Figure 8. (**a**) SEM image of sample GEL02; maps of (**b**) Al, (**c**) Si, (**d**) K, (**e**) Mg, (**f**) Ti and (**g**) Fe.

Table 3. Comparative evaluation on black gloss features (visual appearance, elemental and mineralogical markers) got from different methods (traditional vs. non-destructive and non-invasive techniques) [22].

Production Site	Thickness of the Black-Gloss	Aesthetical Appearance	SR-Based Methods			Other Non-Destructive or Non-Invasive Lab Methods on Micro-Samples		Traditional Destructive Techniques (Literature Data)	
			Elemental Markers in On-Glaze Geometry (SR-XRF)	Elemental Markers in Cross-Section Geometry (SR-XRF)	Mineralogical Composition (SR-XANES)	Elemental Markers in On-Gloss Geometry (Portable XRF)	Elemental Markers in Cross-Section Geometry (SEM-EDS)	Chemical Bulk Tracers (Literature, Benchtop XRF and ICP-MS)	Mineralogical Composition (Literature, XRD)
Attic	~20 µm	Compact and glossy	Fe-based black-gloss, no relevant markers	Black-gloss: major Fe; minor Rb, Zn Clay paste: Ni, Cr	hercynite and magnetite	Major: Fe. Minor: Si, K	Al-silicate K rich gloss. Not relevant markers; sharp interface between gloss and body	Ni, Cr [35]	Quartz, plagioclase, hematite [35]
Laconian	~10 µm	Matte and black, homogeneous	Fe-based black-gloss, no relevant makers	Black-gloss: major Fe, Clay paste: Rb, Ca	hercynite and magnetite	Major: Fe, Mn. Minor: Si, K	Lower Fe, Si and Al amount	No data available	No data available
Central-southern Sicily (Gela colony)	~15 µm	matte and brownish, local reddish	Fe-based black-gloss, no relevant markers	Black-gloss: major Fe, Rb, Mn Clay paste: K	Magnetite and hematite	Major: Fe. Minor: Mn, K	Al-silicate K rich gloss. Not relevant markers. The gloss layer fades out into the body	No data available	No data available
South-Italian colonies (southern Calabria and the Strait of Messina area)	20–40 µm	Matte and compact black/black-bluish gloss	Zn	Black-gloss: major: Zn, Fe Clay paste: (Locrian) Ni, Cr; (Chalcidian) Rb, K, Mn	Hercynite, magnetite (minor) and spinel gahnite	Major: Fe. Minor: Zn, Si, K, Mn	Al-silicate K rich gloss. Not relevant markers. The gloss layer fades out into the body	Chalcidian: Ni, Cr Locrian: Ni, Cr [35]	Chalcidian: quartz, plagioclase, K-feldspar, illite, muscovite, hematite Locrian: quarzt, plagioclase, pyroxene, calcite, iron oxides [35]

4. Discussion

From experimental studies it is known that in order to obtain a good quality black gloss ceramic a multi-phase firing process it is required, with a precise control of temperature and firing duration during each ORO (oxidizing-reducing-oxidizing) phase. These parameters also affect the final appearance of the black gloss, turning from bluish to brownish red [31,45]. Generally speaking, if the slip is «well-made», no hematite or maghemite (Fe^{3+}) is formed and only Fe^{2+} should be present. Otherwise, if Fe^{3+} phases are present, the location of the different Fe-phases inside the gloss is of interest. Fe^{3+} phases usually form a thin slip under/on the top of the black-gloss. When other phases are present—likely the Zn-spinel phases observed in south-Italian products—it is relevant to investigate their mineralogical nature and their location inside the gloss thickness.

The results of this investigation evidenced that in Attic vessels the Fe-based black gloss is mainly due to magnetite and hercynite, indicating an ideal and well-made production routine. On the other hand, in Sicilian colonial vessels (Geloan) prevalently hematite has been found, indicating a different technological signature for the black-gloss, which in fact appears quite different in its aesthetical features (matte and brownish red). As per South-Italian workshops and Laconian products, in some cases only hercynite has been found, indicating well-made pieces. In other cases, Fe^{3+}—diffusely distributed on the gloss thickness—would indicate not perfectly assessed production routines or unstable conditions in ancient kilns.

Overall, South-Italian workshops (Chalcidian and probably Locrian) seem to be discriminated for their Zn signature, indicating a shared fabrication method in these colonies. In fact, as suggested by the literature, Zn could be considered not only a provenance tracer but a technological marker, related to the clay refinement methods [43]. Other studies would correlate the Zn occurrence as typical of specific Greek clays used for the black gloss [46], however, from our characterization studies, such higher Zn traces have to be considered as specific of South-Italian products. XANES spectra and maps collected on these samples revealed the nature of Zn-spinel as $ZnAl_2O_4$ (Gahnite) which is localized over the entire gloss thickness. Experimental studies would be highly useful to better understand the technological choices behind these specific colonial products.

5. Conclusions

Synchrotron X-ray microprobes offers challenging methods for ancient materials characterization studies. In the last years, numerous set-ups have been optimized for CH applications, thus enlarging the current research perspectives.

In ancient ceramics studies, X-ray SR based methods might be able to help answering relevant question related to provenance and technology. Such information can be achieved by combining at least two different methods, with the selection primarily being dependent on (i) the availability of samples/sampling allowance (e.g., entire vessels vs. micro-samples, or cross-sections and microtome samples) (ii) sample preparation and (iii) the length scale required (from the whole clay paste or glaze composition to the detection of crystallites or other textural/composition features into micro or nano layers).

μXRF and μXANES mapping at synchrotron X-ray microprobes can be an ideal combination, especially when no sample preparation is allowed, small fragments are available and a non-destructive approach of the samples is required. Samples can be scanned rapidly in different geometries, for example in cross-section to map elements localization on different layers. The 2D micro-chemical and phase-distribution maps might reveal the location of relevant geochemical tracers—both provenance and technology related—and can be also used for selecting regions of interest in XANES analysis in fluorescence mode. Then, μXANES spectra and maps can be useful in localizing mineral phase distributions and provide relevant information to the better understanding of technological issues in specific ceramic class manufacture. All this work can be carried out by the guidance and screening provided by XRF analyses, which, by being a multi-elemental technique, allows inspecting the samples and determining which sites are the most relevant for μXANES spectra and maps.

The selected case study appears quite meaningful, both in term of archaeological relevance and practical issues. Black gloss ceramics—along with red and black figures vessels—usually include particularly exquisite corpora and rare masterpieces that cannot be sampled or sacrificed for destructive analysis. In addition, classification is still based on typological or style features, thus imposing the development of a new and—hopefully—non-destructive approach for the correct production site identification, especially among the well-known colonial workshops. The combination of XRF and XANES mapping methods with micrometric resolution was able to provide geochemical fingerprints for provenance determination and phase analysis for technological studies, which would benefit by experimental archaeology tests for further information. The comparison with classical microchemical methods pointed out the merits of such an approach, suggesting that even portable X-ray methods might enable fast and preliminary classification useful in defining criteria for micro-sampling operation.

Continuous advances in X-ray microprobes associated with SR source upgrades will allow for even more powerful imaging and spectroscopic methods. In the coming future, the third and fourth generation SR sources will offer increasing opportunities for analyzing tiny and complex objects, which are typical among cultural heritage materials, in a faster way and with higher resolutions.

Supplementary Materials: The following are available online at https://www.mdpi.com/2076-3417/11/17/8052/s1, Figure S1: portable XRF maps acquired by Elio ©Bruker. Figure S2: Fe-XANES point spectra collected on the gloss in cross-section geometry for GEVN (a), GEL (b) and ME (c) samples. The spectra were collected on a 3 μm × 5 μm area, across Fe K-edge, as detailed in materials and methods. Figure S3: Fe-XANES spectra of Fe^{2+} and Fe^{3+} standard with the indication of the two energies E1 (7127 eV) and E2 (7132 eV), used to differentiate the contribution of the two phases in the XRF Fe XANES maps shown in Figure 3. Figure S4: Zn K-edge XANES reference spectra from [43] where we indicated the energies chosen for the Zn XANES maps, later used to identify the predominant phase. Figure S5: Si/Al and Fe(Si+Al) tenors in a selection of black gloss representative of the different identified products.

Author Contributions: Conceptualization, A.G. and S.R.; methodology, A.G., S.R. and S.S.; software, A.G. and S.R.; validation, A.G., S.R. and M.L.; formal analysis, A.G., G.K. and S.R.; investigation, A.G., S.R., M.L. and S.S.; resources, A.G., S.R. and S.S.; data curation, A.G., G.K. and S.R.; writing—original draft preparation, A.G. and S.R.; writing—review and editing, A.G., S.S., A.S., G.B., P.M., S.R. and M.L.; visualization, A.G., S.S., A.S., G.B., P.M. and S.R.; supervision, A.G. and S.R.; project administration, A.G.; funding acquisition, A.G. All authors have read and agreed to the published version of the manuscript.

Funding: This research received no external funding.

Institutional Review Board Statement: Not applicable.

Informed Consent Statement: Not applicable.

Data Availability Statement: The data presented in this study are available on request from the corresponding author. The data are not publicly available due to privacy reasons including an embargo period according to the facility's scientific data policy.

Acknowledgments: The research leading to these results has received a financial support in the framework of the EU CALIPSOplus TransNational Access programme. We acknowledge SOLEIL for provision of synchrotron radiation facilities.

Conflicts of Interest: The authors declare no conflict of interest.

References

1. Shackley, M.S. (Ed.) *X-ray Fluorescence Spectrometry (XRF) in Geoarchaeology*; Springer: New York, NY, USA, 2011; ISBN 978-1-4419-6885-2.
2. Mantler, M.; Schreiner, M. X-ray Fluorescence Spectrometry in Art and Archaeology. *X-ray Spectrom.* **2000**, *29*, 3–17. [CrossRef]
3. Hall, M.E. X-ray Fluorescence-Energy Dispersive (ED-XRF) and Wavelength Dispersive (WD-XRF) Spectrometry. In *The Oxford Handbook of Archaeological Ceramic Analysis*; Hunt, A., Ed.; Oxford University Press: Oxford, UK, 2017; ISBN 978-0-19-968153-2.
4. Maritan, L. Archaeo-Ceramic 2.0: Investigating Ancient Ceramics Using Modern Technological Approaches. *Archaeol. Anthropol. Sci.* **2019**, *11*, 5085–5093. [CrossRef]
5. Hein, A.; Tsolakidou, A.; Iliopoulos, I.; Mommsen, H.; Buxeda i Garrigós, J.; Montana, G.; Kilikoglou, V. Standardisation of Elemental Analytical Techniques Applied to Provenance Studies of Archaeological Ceramics: An Inter Laboratory Calibration Study. *Analyst* **2002**, *127*, 542–553. [CrossRef]
6. Waksman, Y. Provenance Studies: Productions and Compositional Groups. In *The Oxford Handbook of Archaeological Ceramic Analysis*; Hunt, A., Ed.; Oxford University Press: Oxford, UK, 2015; pp. 1–18.
7. Barone, G.; Mazzoleni, P.; Spagnolo, G.V.; Raneri, S. Artificial Neural Network for the Provenance Study of Archaeological Ceramics Using Clay Sediment Database. *J. Cult. Herit.* **2019**, *38*, 147–157. [CrossRef]
8. De Bonis, A.; Arienzo, I.; D'Antonio, M.; Franciosi, L.; Germinario, C.; Grifa, C.; Guarino, V.; Langella, A.; Morra, V. Sr-Nd Isotopic Fingerprinting as a Tool for Ceramic Provenance: Its Application on Raw Materials, Ceramic Replicas and Ancient Pottery. *J. Archaeol. Sci.* **2018**, *94*, 51–59. [CrossRef]
9. Renson, V.; Slane, K.W.; Rautman, M.L.; Kidd, B.; Guthrie, J.; Glascock, M.D. Pottery Provenance in the Eastern Mediterranean Using Lead Isotopes. *Archaeometry* **2016**, *58*, 54–67. [CrossRef]
10. Vandenabeele, P.; Donais, M.K. Mobile Spectroscopic Instrumentation in Archaeometry Research. *Appl. Spectrosc.* **2016**, *70*, 27–41. [CrossRef] [PubMed]
11. Holmqvist, E. Handheld Portable Energy-Dispersive X-ray Fluorescence Spectrometry (pXRF). In *The Oxford Handbook of Archaeological Ceramic Analysis*; Hunt, A., Ed.; Oxford University Press: Oxford, UK, 2015; pp. 1–23.
12. Gianoncelli, A.; Castaing, J.; Ortega, L.; Dooryhée, E.; Salomon, J.; Walter, P.; Hodeau, J.-L.; Bordet, P. A Portable Instrument for in Situ Determination of the Chemical and Phase Compositions of Cultural Heritage Objects. *X-ray Spectrom.* **2008**, *37*, 418–423. [CrossRef]
13. Hunt, A.M.W.; Speakman, R.J. Portable XRF Analysis of Archaeological Sediments and Ceramics. *J. Archaeol. Sci.* **2015**, *53*, 626–638. [CrossRef]
14. Alfeld, M.; Pedroso, J.V.; van Eikema Hommes, M.; Van der Snickt, G.; Tauber, G.; Blaas, J.; Haschke, M.; Erler, K.; Dik, J.; Janssens, K. A Mobile Instrument for in Situ Scanning Macro-XRF Investigation of Historical Paintings. *J. Anal. At. Spectrom.* **2013**, *28*, 760–767. [CrossRef]
15. Romano, F.P.; Caliri, C.; Cosentino, L.; Gammino, S.; Giuntini, L.; Mascali, D.; Neri, L.; Pappalardo, L.; Rizzo, F.; Taccetti, F. Macro and Micro Full Field X-ray Fluorescence with an X-ray Pinhole Camera Presenting High Energy and High Spatial Resolution. *Anal. Chem.* **2014**, *86*, 10892–10899. [CrossRef]
16. Romano, F.P.; Janssens, K. Preface to the Special Issue on: MA-XRF "Developments and Applications of Macro-XRF in Conservation, Art, and Archeology" (Trieste, Italy, 24 and 25 September 2017). *X-ray Spectrom.* **2019**, *48*, 249–250. [CrossRef]
17. Janssens, K.; Cotte, M. Using Synchrotron Radiation for Characterization of Cultural Heritage Materials. In *Synchrotron Light Sources and Free-Electron Lasers: Accelerator Physics, Instrumentation and Science Applications*; Jaeschke, E.J., Khan, S., Schneider, J.R., Hastings, J.B., Eds.; Springer International Publishing: Cham, Switzerland, 2020; pp. 2457–2483. ISBN 978-3-030-23201-6.
18. Bertrand, L.; Robinet, L.; Thoury, M.; Janssens, K.; Cohen, S.X.; Schöder, S. Cultural Heritage and Archaeology Materials Studied by Synchrotron Spectroscopy and Imaging. *Appl. Phys. A* **2012**, *106*, 377–396. [CrossRef]
19. Adams, F.; Janssens, K.; Snigirev, A. Microscopic X-ray Fluorescence Analysis and Related Methods with Laboratory and Synchrotron Radiation Sources. *J. Anal. At. Spectrom.* **1998**, *13*, 319–331. [CrossRef]
20. Cotte, M.; Genty-Vincent, A.; Janssens, K.; Susini, J. Applications of Synchrotron X-ray Nano-Probes in the Field of Cultural Heritage. *C. R. Phys.* **2018**, *19*, 575–588. [CrossRef]
21. Cotte, M.; Pouyet, E.; Salomé, M.; Rivard, C.; Nolf, W.D.; Castillo-Michel, H.; Fabris, T.; Monico, L.; Janssens, K.; Wang, T.; et al. The ID21 X-ray and Infrared Microscopy Beamline at the ESRF: Status and Recent Applications to Artistic Materials. *J. Anal. At. Spectrom.* **2017**, *32*, 477–493. [CrossRef]
22. Gianoncelli, A.; Raneri, S.; Schoeder, S.; Okbinoglu, T.; Barone, G.; Santostefano, A.; Mazzoleni, P. Synchrotron M-XRF Imaging and μXANES of Black-Glazed Wares at the PUMA Beamline: Insights on Technological Markers for Colonial Productions. *Microchem. J.* **2020**, *154*, 104629. [CrossRef]
23. Bertrand, L.; Cohen, S.X.; Thoury, M.; David, S.; Schoeder, S. IPANEMA, Un Laboratoire Dédié à l'étude Des Matériaux Anciens et Patrimoniaux Par Méthodes Synchrotron. *Reflets Phys.* **2019**, *63*, 21. [CrossRef]
24. Sciau, P.; Goudeau, P.; Tamura, N.; Dooryhee, E. Micro Scanning X-ray Diffraction Study of Gallo-Roman Terra Sigillata Ceramics. *Appl. Phys. A* **2006**, *83*, 219–224. [CrossRef]
25. Gliozzo, E.; Kirkman, I.W.; Pantos, E.; Turbanti, I.M. Black Gloss Pottery: Production Sites and Technology in Northern Etruria, Part II: Gloss Technology. *Archaeometry* **2004**, *46*, 227–246. [CrossRef]
26. Pradell, T.; Molera, J.; Salvadó, N.; Labrador, A. Synchrotron Radiation Micro-XRD in the Study of Glaze Technology. *Appl. Phys. A* **2010**, *99*, 407–417. [CrossRef]

27. Farges, F.; Cotte, M. X-ray Absorption Spectroscopy and Cultural Heritage: Highlights and Perspectives. In *X-ray Absorption and X-ray Emission Spectroscopy*; van Bokhoven, J.A., Lamberti, C., Eds.; John Wiley & Sons, Ltd.: Hoboken, NJ, USA, 2016; pp. 609–636. ISBN 978-1-118-84424-3.

28. Sciau, P.; Wang, T. *Full-Field Transmission X-ray Microspectroscopy (FF-XANES) Applied to Cultural Heritage Materials: The Case of Ancient Ceramics*; IntechOpen: London, UK, 2019; ISBN 978-1-83880-442-8.

29. Sciau, P.; Leon, Y.; Goudeau, P.; Fakra, S.C.; Webb, S.; Mehta, A. Reverse Engineering the Ancient Ceramic Technology Based on X-ray Fluorescence Spectromicroscopy. *J. Anal. At. Spectrom.* **2011**, *26*, 969–976. [CrossRef]

30. Meirer, F.; Liu, Y.; Pouyet, E.; Fayard, B.; Cotte, M.; Sanchez, C.; Andrews, J.C.; Mehta, A.; Sciau, P. Full-Field XANES Analysis of Roman Ceramics to Estimate Firing Conditions—A Novel Probe to Study Hierarchical Heterogeneous Materials. *J. Anal. At. Spectrom.* **2013**, *28*, 1870–1883. [CrossRef]

31. Cianchetta, I.; Maish, J.; Saunders, D.; Walton, M.; Mehta, A.; Foran, B.; Trentelman, K. Investigating the Firing Protocol of Athenian Pottery Production: A Raman Study of Replicate and Ancient Sherds. *J. Raman Spectrosc.* **2015**, *46*, 996–1002. [CrossRef]

32. Cianchetta, I.; Trentelman, K.; Maish, J.; Saunders, D.; Foran, B.; Walton, M.; Sciau, P.; Wang, T.; Pouyet, E.; Cotte, M.; et al. Evidence for an Unorthodox Firing Sequence Employed by the Berlin Painter: Deciphering Ancient Ceramic Firing Conditions through High-Resolution Material Characterization and Replication. *J. Anal. At. Spectrom.* **2015**, *30*, 666–676. [CrossRef]

33. Bertrand, L.; Cotte, M.; Stampanoni, M.; Thoury, M.; Marone, F.; Schöeder, S. Development and Trends in Synchrotron Studies of Ancient and Historical Materials. *Phys. Rep.* **2012**, *519*, 51–96. [CrossRef]

34. Iozzo, M. *Ceramica "Calcidese". Nuovi Documenti e Problemi Riproposti. Atti e Memorie della Società Magna Grecia, III s., II (1993)*; John Wiley & Sons, Inc.: Roma, Italy, 1994.

35. Barone, G.; Ioppolo, S.; Majolino, D.; Branca, C.; Sannino, L.; Spagnolo, G.; Tigano, G. Archaeometric Analyses on Pottery from Archaeological Excavations in Messina (Sicily, Italy) from the Greek Archaic to the Medieval Age. *Period. Mineral.* **2005**, *74*, 11–41.

36. Tigano, G. IsolatoS. Via Industriale. Lo scavo e i primi dati sui materiali. In *Da Zancle a Messina. Un Percorso Archeologico Attraverso Gli Scavi, I*; Bacci, G.M., Tigano, G., Eds.; Sicania: Palermo, Italy, 1999; pp. 123–155.

37. Barone, G.; Ioppolo, S.; Majolino, D.; Migliardo, D.; Sannino, L.; Spagnolo, G.; Tigano, G. Contributo allo studio delle ceramiche provenienti dagli scavi di Messina. Risultati preliminari. In *Da Zancle a Messina. Un Percorso Archeologico Attraverso Gli Scavi, II*; Bacci, G.M., Tigano, G., Eds.; Sicania: Palermo, Italy, 1999; pp. 87–117.

38. Stibbe, C.M. Laconian mixing-bowls: A history of the krater lakonikos from the 7th to the 5th century B.C. In *Laconian Black-Glazed Pottery I*; Allard Pierson Museum: Amsterdam, The Netherlands, 1989.

39. Stibbe, C.M. Laconian drinking vessels and other open shapes. In *Laconian Black-Glazed Pottery II*; Allard Pierson Museum: Amsterdam, The Netherlands, 1994.

40. Mirti, P.; Casoli, A. Analysis and classification of ceramic material excavated on a South Italian archaeological site. *Ann. Chim.* **1995**, *85*, 519–530.

41. Mirti, P.; Gulmini, M.; Pace, M.; Elia, D. The provenance of red figured vases from Locri Epizephiri (Southern Italy): New evidence by chemical analysis. *Archaeometry* **2004**, *46*, 183–200. [CrossRef]

42. Bertrand, L.; Languille, M.-A.; Cohen, S.X.; Robinet, L.; Gervais, C.; Leroy, S.; Bernard, D.; Le Pennec, E.; Josse, W.; Doucet, J.; et al. European Research Platform IPANEMA at the SOLEIL Synchrotron for Ancient and Historical Materials. *J. Synchrotron Radiat.* **2011**, *18*, 765–772. [CrossRef]

43. Walton, M.; Trentelman, K.; Cianchetta, I.; Maish, J.; Saunders, D.; Foran, B.; Mehta, A. Zn in Athenian Black Gloss Ceramic Slips: A Trace Element Marker for Fabrication Technology. *J. Am. Ceram. Soc.* **2015**, *98*, 430–436. [CrossRef]

44. Tack, P.; Bazi, B.; Vekemans, B.; Okbinoglu, T.; Van Maldeghem, F.; Goderis, S.; Schöder, S.; Vincze, L. Investigation of (Micro) Meteoritic Materials at the New Hard X-ray Imaging PUMA Beamline for Heritage Sciences. *J. Synchrotron Radiat.* **2019**, *26*, 2033–2039. [CrossRef] [PubMed]

45. Aloupi-Siotis, E. Ceramic Technology: How to Characterise Black Fe-Based Glass-Ceramic Coatings. *Archaeol. Anthropol. Sci.* **2020**, *12*, 191. [CrossRef]

46. Chaviara, A.; Aloupi-Siotis, E. The Story of a Soil That Became a Glaze: Chemical and Microscopic Fingerprints on the Attic Vases. *J. Archaeol. Sci. Rep.* **2016**, *7*, 510–518. [CrossRef]

MDPI

St. Alban-Anlage 66

4052 Basel

Switzerland

Tel. +41 61 683 77 34

Fax +41 61 302 89 18

www.mdpi.com

Applied Sciences Editorial Office

E-mail: applsci@mdpi.com

www.mdpi.com/journal/applsci

* 9 7 8 3 0 3 6 5 4 8 6 7 8 *